STEPHENSON'S
ROCKET

1829 onwards

COVER IMAGE: **A profile view of Stephenson's**
Rocket. *(John Glithero)*

First published in June 2016

A catalogue record for this book is available
from the British Library

ISBN 978 1 78521 063 1

Library of Congress control no. 2015948111

Design and layout by James Robertson

Published by Haynes Publishing,
Sparkford, Yeovil,
Somerset BA22 7JJ, UK.
Tel: 01963 440635
Int. tel: +44 1963 440635
Website: www.haynes.co.uk

Haynes North America Inc.,
861 Lawrence Drive, Newbury Park,
California 91320, USA.

Printed in the USA by Odcombe Press LP,
1299 Bridgestone Parkway, La Vergne,
TN 37086.

Science Museum logo © SCMG Enterprises Ltd.

This book is produced in association with the
Science Museum, London.
Sales of this book help support the museum's
exhibitions and programmes.
www.sciencemuseum.org.uk

All images © Science & Society Picture Library,
unless otherwise stated.

Acknowledgements

This book could not have been delivered without the assistance of many
people who have willingly given me help, for which I am very grateful.

I am a seat-of-the-pants mechanical engineer who cares
passionately about the subject and loves sharing that passion with
people! However, I had no idea how much help I would need to
make sense of *Rocket*'s story in order to do it justice and to deliver
the contents of this Haynes Manual on time. I am grateful to all those
who provided that help.

Key players informing the production of the book were Dr Michael
Bailey and John Glithero, who unlocked the magic by doing all the
hard work in 1999 with their original research into *Rocket*. They
produced the definitive publication *The Engineering and History of
Rocket* and have given me their help and blessing in writing this book.
Many of their findings, fine CAD drawings and photographs come
from that publication, courtesy of the Science and Society Picture
Library. Significant National Railway Museum help has come from
Sarah Norville in particular, but also John Clarke, Ed Bartholomew
and Peter Heaton with access to the National Railway Museum's
records, photographs and collections. Special thanks to Adrian
Lucey for the detail photographs of the sectioned National Railway
Museum replica and great technical help collating the images. Also
Dave Burrows for creating and making available the National Railway
Museum's Workshop Notebook file. But many other people also
need to be thanked: Lynn Patrick, Martyn Stevens, David Mosley,
Robert Sjoo, Jim Rees, Steve Rendle, Bill Parker, Geoff Phelps,
Graham Morris, Andrew Scott, Nev Cumberland, Fred Bailey, David
Moore, Crispin Cousins, Charlie Bird, Dai Jones, Steve Thorpe, Dave
Campey and John Chambers. Finally the most important thanks
go to my wife, Angela, for not only putting up with me during the
cloistered period of writing, but for her critical reading of the book
suggesting many amendments to make the wording more concise,
less technical and more accessible. She is not a technical person but
we reasoned that if she could understand the contents of the manual,
the *Rocket* story would be accessible to a general readership.

Richard Gibbon OBE
York, March 2016

STEPHENSON'S ROCKET

1829 onwards

Owners' Workshop Manual

An insight into the design, construction, operation and
maintenance of the iconic steam locomotive

Richard Gibbon OBE

Contents

6 Introduction

12 The *Rocket* story

The Liverpool and Manchester Railway   14
The challenge of the track system   17
The challenge of the boiler   18
The Rainhill Trials proposed   21
Rocket is conceived   22
Putting the challenge on the line   22
The Rainhill Trials   25
Rocket after Rainhill   29
The opening of the Liverpool and
     Manchester Railway   32
Rocket after the Liverpool and
     Manchester Railway opening   34
Rocket as a museum exhibit   35
What happened to the also-rans
     after the Rainhill Trials?   40

42 The anatomy of *Rocket*

Introduction: *Rocket*'s vital statistics   44
The basic design   44
Tractive effort   47
The frames   48
The cylinders   50
The wheels and suspension   52
The motion   54
The boiler and related systems   56
The water gauges   60
The feed pump   61
The safety valves   62
The dome   63
The regulator   64
The grate   64
The tender   65
The footplate and fittings   67
The paintwork   68

70 Driving, firing and riding on *Rocket*

Driving, firing and riding with the
     original *Rocket*   72
Fanny Kemble's account   72
Thomas Creevey's account   73
A fireman's account   74
Edward Entwhistle's account   78
An account of driving a *Rocket* replica   81

82 | Maintaining *Rocket*

Scheduled maintenance programme   84
Water gauge maintenance   85
Soot and ash deposits   86
Water quality and the effectiveness
  of the boiler   86
Corrosion   88
Lubrication   89
Bearing adjustment   90
Glands on sliding joints   91
Grate   93
Tender water barrel   93
Wheel wear   94
Conclusion   95

96 | *Rocket* replicas

The concept of replicas   98
The 1881 Crewe replica   98
The Henry Ford, R. Stephenson & Co
  replicas   101
The 1923 Buster Keaton movie replica   102
The 1937 rebuilt Crewe replica for the LMS 102
The 1979 Mike Satow replica   103
The 2010 Jim Rees replica   103
The function of replicas and what we
  learn from them   103

108 | The value of running with replicas

Replicas that operate   110
The Beamish collection   110
The 2002 BBC *Timewatch* recreation of
  the Rainhill Trials   111
Preparing the *Novelty* replica to run   112
The re-enactment of the Rainhill Trials 116
The 2010 replacement *Rocket* replica   120
Cautionary tales!   124
Rocket and *The Day the World Took Off*
  filming   126

128 | Epilogue: the legacy of *Rocket*

Introduction   130
The industrial legacy   130
The social legacy   134
The cultural legacy   136
The technological and engineering legacy 138
A personal legacy   138

142 | Appendices

1 The significance of the Dr Michael
    Bailey and John Glithero report   144
2 Glossary of some terms used in
    this manual   148
3 *Rocket*'s working life timeline   150
4 Relevant places to visit   152

154 | Index

Introduction

If you were to ask a child to name four famous locomotives or trains, my bet would be that Thomas the Tank Engine, Stephenson's *Rocket*, the Bullet Train and perhaps *Flying Scotsman* would be amongst the answers. The unusual name, shape and colour of Stephenson's *Rocket* have etched themselves into the consciousness of the British nation. This is the story of the machine that won hearts and minds during the early years of steam locomotion, as well as an explanation of this extraordinary engineering achievement.

I was lucky enough to be able to work in an organisation that gave me unique opportunities to be intimately involved with the design of Stephenson's original *Rocket*. As chief engineer of the National Railway Museum in York between 1989 and 2003, I learned so much about the extant remains of *Rocket*. I was also involved closely with the operating replica built by Mike Satow for the Liverpool & Manchester Railway (L&MR) anniversary celebrations in 1980 held at Rainhill and known as *Rocket 150*. This replica then went on to travel the world as a roving ambassador for the National Railway Museum.

Most significantly, in 1999 the once in a lifetime opportunity arose on my watch to intercept *Rocket* upon its return from display at a UK-inspired exhibition in Japan, and

BELOW *Rocket* **as built in 1829. This drawing was made by the Science Museum in 1950 and represents** *Rocket* **in its Rainhill Trials condition.**

bring it to the National Railway Museum for experts to carry out a controlled archaeological examination and explore the true story of its engineering genesis and development. The locomotive was destined to take centre stage in the Science Museum's new exhibition called 'Making the Modern World' in 2000, but *Rocket*'s place had not then been fully prepared. That left us at the National Railway Museum with an opportunity not only to have *Rocket* on display in York temporarily but also to use that time to further public knowledge about the locomotive by getting experts to carefully strip, examine and then rebuild it in the public arena.

We were fortunate that two world-renowned experts on early locomotive engineering formed the team for the 'dig'. Much of what the reader will find in this book would have lain undiscovered were it not for the significant work of Dr Michael Bailey and Dr John Glithero, working on the floor space of the National Railway Museum during 1999. The evidence was published in a great work that

ABOVE *Rocket* **nameplate. This is a remarkable survival, considering that most of the non-ferrous material from** *Rocket* **was removed and lost when the locomotive was in storage at Robert Stephenson & Co.**

BELOW *Rocket* **1979 replica at the** *Rocket 150* **cavalcade, 1980. The wooden wheels of the replica are visible here and the wheels are only just on the line – a portent of trouble in store. The replica has been fitted with a temporary front buffer beam to enable it to be shunted with other vehicles.**

built for British Railways in 1960, there have been many significant, historic and special steam locomotives amongst the hundreds of thousands built for worldwide railway use. However, there is no doubt in my mind that *Rocket* was the locomotive that changed the world of railway engineering. It was radically different from anything that had gone before and yet it set the pattern and principles for the building of future machines that remained substantially unchanged for 130 years.

This book looks at the features and arrangement of *Rocket* and attempts to analyse why this machine took the world of railways by storm! *Why* was it so special? Surely there are other machines that radically changed the way we did things thereafter? I have looked for comparable stories from the world of engineering that match the magnitude of the *Rocket* achievement. The mobile phone, the internal combustion engine, the pneumatic tyre, the ball-point pen, the air blade hand-drier, the jet engine perhaps; but unlike these inventions, *Rocket* was an inspired rearrangement of the features that existed on previous locomotives. It was the way those features were combined that gave it a flying start. What George and Robert Stephenson did so spectacularly was to assemble the parts in a way that not only fulfilled the written specification but also brought

ABOVE Dr Michael Bailey and Dr John Glithero at work on the 'dig' in 1999.

forms the definitive *Rocket* textbook, called *The Engineering and History of 'Rocket'*, written by Bailey and Glithero and published by the Science Museum in 2000.

From the start of Richard Trevithick's crude 1804 *Pen y Darren* locomotive's clumsy first movements in South Wales to the withdrawal from service of the last steam locomotive

RIGHT Three veterans built in 1829 share the limelight at the National Railway Museum, 1999: *Rocket*, with Bailey and Glithero in attendance, is accompanied by *Agenoria* (left) and *Sans Pareil* (right).

the components together into a critical mass. This was what made all subsequent makers want to replicate their design principles – despite its design being rubbished by critics at the time of its launch in 1829!

Steam locomotives prior to *Rocket*'s debut at the Rainhill Trials of 1829 were more or less ingenious contraptions that clever artisans put together to do a specific job adequately. It might seem callous to write off those early machines as *adequate*, but haulage on rails using horses and stationary steam engines was the norm for wheeled transport in the 19th century. Steam locomotion was new and exceptional at a time when the Industrial Revolution was in full steam throughout

ABOVE **Richard Trevithick's first steam locomotive to run on rails was built at Coalbrookdale in Shropshire and operated in 1803. A similar locomotive ran at Pen y Darren in Wales, hauling a mineral train in 1804 carrying ten tons of iron and 70 men a distance of ten miles in four hours before exploding.**

LEFT *Evening Star*, **the last steam locomotive to be built for British Railways, embodied all the original features of** *Rocket*, **built 130 years earlier – the multi-tubular boiler, the blast pipe and the direct drive to the cylinders.**

RIGHT George Stephenson's Liverpool and Manchester days. Behind this painting of George can be seen an early Robert Stephenson locomotive hauling a train across the treacherous bog at Chat Moss. A drainage ditch is shown to the right. Crossing this bog successfully was one of George's greatest achievements.

Britain. What Robert Stephenson and his father did so cleverly with *Rocket* was to point the way for self-propelled motion to transport goods and passengers.

Although this book is about *Rocket* and its story, the tale of the locomotive is inextricably linked with the extraordinary life of George Stephenson and his son Robert, who was to go on to become one of the world's most famous civil engineers. George had humble Northumbrian roots and grew up in a large family, showing a genius for all things mechanical even though he was unable to read or write. His father could not afford to send him to school. At 18 George enrolled for night school lessons from a local schoolmaster. Throughout his early working life as a pit labourer at Willington, and later as a colliery engineer at Killingworth seven miles north of Newcastle, he was able to bring about improvements in many of the machines with which he came into contact, whether they were pit steam engines or clocks. He had a natural ability to understand what was going on inside complex mechanisms and to repair or improve them. He was a seat-of-the-pants engineer who understood machines instinctively and called a spade a spade! His straight speaking and strong Northumbrian accent meant that there were many people in the field of mechanical engineering, as well as politicians, who could not accept that he was a genius.

At 21 he married Frances Henderson. Their only child Robert was born in 1803. Tragically Frances died of tuberculosis, leaving George as a single parent of a three-year-old boy. The close bond that developed between them over the ensuing years developed into a dynamic combination of engineering genius and determination. When Robert came home from school as a young man, he spent the evenings teaching his father what he had

LEFT Conjectural painting of George Stephenson. He is shown at Killingworth waggon-way surrounded by his family and his achievements, including the 'Geordie' miners' lamp and one of his early steam locomotives. His family house is in the background.

learned. Robert later served an apprenticeship in mine surveying under Nicholas Wood (of whom more later). When Robert was 19 he took up a place at Edinburgh University to study natural philosophy, natural history and chemistry. It is significant that while there he won a prize and chose to receive Charles Hutton's *Tracts on Mathematical and Philosophical Subjects*, the first seven chapters of which related to *The Principles of Bridges*. This was a portent of what was to come later in his life, as he became one of the greatest civil engineers the world has known. What a remarkable achievement for his father to have nurtured such a genius in such difficult personal and domestic circumstances! George and Robert working together made an unstoppable combination of genius and practical engineering knowledge that peaked with their creation of *Rocket*.

I am thrilled to have been commissioned to tell the story of *Rocket*. It is hard to believe when looking at the much changed and rather dull incomplete remains of *Rocket* standing in the Science Museum that this once vibrant and attractive machine stirred public hearts and minds to the point where it must have felt like the Moon-landing of its day! *Rocket* still takes centre stage in the exhibition called 'The Making the Modern World' at South Kensington's Science Museum in London.

This book is laid out in the standard Haynes format, with the story of *Rocket* followed by chapters on its anatomy and what it was like to drive and maintain. I have introduced a section on the replica locomotives and running the replicas, as this has been such an important part of communicating the story of *Rocket* to a wide public audience throughout the world. The book closes with an epilogue that examines the legacy of *Rocket*.

Images for this publication have been drawn from many sources, but principally from the Science and Society Picture Library (SSPL) based at the Science Museum. Where sources are not credited, they belong to SSPL.

LEFT *Rocket* remains, as displayed in the 'Making the Modern World' exhibition at South Kensington, London. The locomotive is shown in its surviving form with large smokebox and cylinders lowered to nearly horizontal. Note the Birkinshaw track with the 'fish-bellied' shape between each sleeper.

Rocket arrives at the National Railway Museum in 1999. The Dr Michael Bailey and Dr John Glithero investigation was carried out in the Great Hall in full public view.

cket
that started a revolution

1829

The Rainhill
Trials

1829

CHAPTER ONE
The *Rocket* story

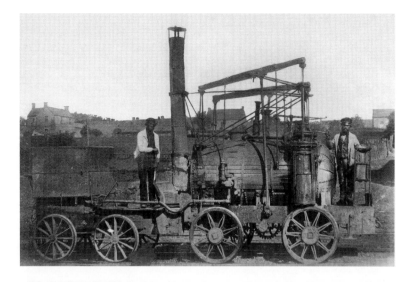

LEFT *Puffing Billy* by William Hedley, 1814, built for Wylam Colliery. The remains of this ancient locomotive are displayed alongside the Stephensons' *Rocket* for comparison in the 'Making the Modern World' exhibition.

The Liverpool and Manchester Railway

Steam locomotives were being put together and operated by various engineers for 25 years before *Rocket* came on the scene in 1829. Examples were *Puffing Billy* in 1814, *Locomotion* in 1825 and *Salamanca* in 1815 (see images on left).

The principle on which steam locomotives worked was that a coal or coke fire burned inside the boiler of the locomotive, part filled with water. The boiler was like a kettle with the lid firmly sealed down. As the water boiled, the upper part of the boiler filled with steam under modest pressure. This steam was led to double-acting pistons, working inside cylinders, which pushed and pulled the wheels around as in a child's pedal car. There were several versions in the early 19th century of how this should be done to best effect. These early locomotives were designed to pull heavy loads on colliery railways at steady speeds.

George Stephenson and his son Robert were both closely involved with the creative process that led to the building of *Rocket*. George had created several locomotives in the previous 15 years and they had operated moderately

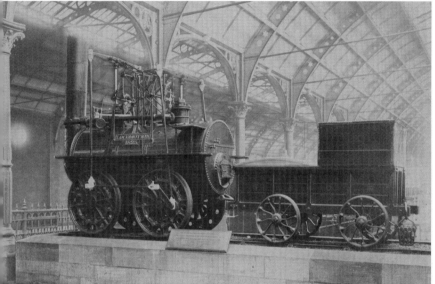

CENTRE *Locomotion*, built by George Stephenson in 1825 for the opening of the Stockton & Darlington Railway. Seen here preserved after a 30-year life, it is shown on a plinth at Darlington Bank Top station. It is now part of the National Collection and can be seen at Head of Steam, Darlington Railway Museum.

LEFT A costume painting entitled *The Collier 1814* shows Blenkinsop's *Salamanca*, built for the Middleton Railway in Leeds, which ran with a rack and pinion to make sure of rail/wheel grip. It shows the value of artwork not primarily intended to record the locomotive.

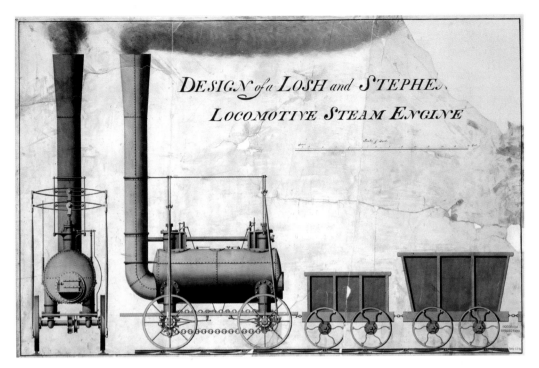

successfully carrying goods over the lines for which they were built. These railways were isolated routes, like those at Middleton in Leeds or Wylam in Northumbria, where the sole purpose of the railway was to enable coal that had been brought up from the mines to be transported to the place where it could be shipped to its final destination.

The Liverpool and Manchester Railway (L&MR) with which George Stephenson was deeply involved in 1828 as engineer and surveyor was, however, a different proposition from those which had gone before. This was

to be a railway carrying goods, but it was also to be the world's first intercity railway carrying passengers. The prospect of a large number of business people wanting to travel regularly between Liverpool and Manchester was held out as a potential source of profit to the developers. From the start this was to be a passenger and goods railway system.

Now, the topography of the chosen route was mostly level, but with some steeply graded sections. Most engineers had assumed that steam locomotives would work the flat portions

RIGHT Map of route for Liverpool and Manchester Railway. Noteworthy (travelling west to east) are Edge Hill, Rainhill Levels, Parkside and Chat Moss and Eccles.

BELOW Gradient profile for Liverpool and Manchester Railway. Noteworthy is the steep hill between Liverpool and Edge Hill, which explains why the passenger railway stopped at Crown Street. Also note the climb up to and down from the Rainhill Levels.

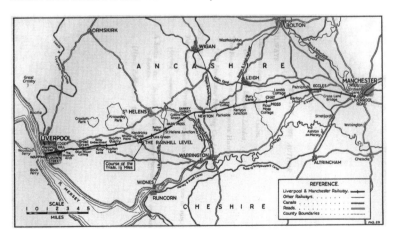

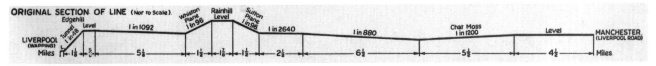

ABOVE Canterbury and
Whitstable Railway built
by George Stephenson
1825–30 with stationary
winding engines. The
rope hauling the train
can be seen between
the tracks.

of the railway and stationary steam winding engines would work the trains up and down the inclines. However, there was doubt in the minds of the directors of the L&MR in 1828 as to whether a steam locomotive was sufficiently advanced to be a reliable haulage system even along the flat sections. There must even have been doubt in George Stephenson's mind. He was also engineer to the Canterbury and Whitstable proposed railway, which gained assent in 1825 and relied on winding engines for steep sections of the track. That was much steeper than the Liverpool and Manchester

route. But there were still hills to climb out of Liverpool and Stephenson knew from his experience at Canterbury that attaching trains to and detaching them from ropes, for the proposed route for the L&MR, would increase the journey time.

The Stephensons faced two other major challenges. Firstly, for their locomotives to succeed on new railways they were being asked to build they would have to solve the problem of creating a track system that could support the new locomotives reliably. Secondly, they needed to design an effective lightweight steam producer,

RIGHT 'Little Eaton
Gangway', as the
Derby Canal Railway
(1795–1908) was
known, showing
cast iron angled rails
suitable for this horse-
drawn line, where the
horse's weight is not
felt by the track.

using a new type of boiler that sustained a high steam production within a compact space.

The challenge of the track system

The first challenge was about the strength of the rails on which the loaded trains ran. A horse-drawn railway merely had to carry the weight of the loaded waggon, as the motive power method of a horse allowed the beast's weight to fall between the tracks where it walked. The early preferences for track design had been for short cast-iron fish-bellied rails resting on stone sleepers set in the ground. A typical loaded four-wheeled waggon might weigh in at three tons, but nearly all of the effective early locomotives before *Rocket*, like William Hedley's *Puffing Billy* of 1814 at Wylam or George Stephenson's *Locomotion* of 1825 built for the Stockton and Darlington Railway, at seven tons weighed much more than this. Cast iron rails were brittle, and although adequate for carrying up to one ton per wheel tended to snap when loaded with a heavy locomotive like *Locomotion*.

To overcome this technical challenge George Stephenson worked with John Birkinshaw of Bedlington Iron Works to produce what they called 'malleable iron' rails. This radical improvement to the rail system was exceptionally clever. Wrought or malleable iron (which was made by a labour-intensive process called 'puddling'), would bend rather

ABOVE *Royal George*, October 1827. This heavy, six-wheeled engine built by Timothy Hackworth showed the way that locomotives were developing at that time. The Stephensons needed to break this mould in order to return to lightweight locomotives. *(Author)*

than snap when overloaded – rather like mild steel nowadays – but it could only be produced in small-quantity batches, typically football-sized. The clever concept of the Stephenson–Birkinshaw malleable rail was to roll out a T-section from the football-sized, red-hot wrought iron lump that spanned six sleepers. The final rolling pass in the manufacturing process was carried out with an eccentric roller that squeezed only the five vertical portions of the T-section that lay *between* the sleeper supports, making a continuous five-length section of fish-belly shaped rail (see below).

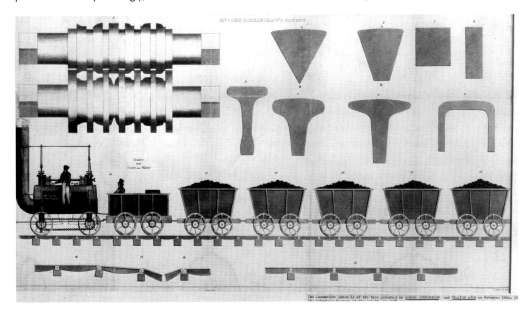

LEFT Design for improved rails in 1824, showing how Birkinshaw Rails were made from wrought iron with a fish-bellied shape between the supports. Each piece spanned six sleepers. The locomotive is the Stephenson/Losh design shown at the top of page 15.

ABOVE *Locomotion* of 1825 on a plinth at Darlington, showing low central fire-tube. The fire only burned in the top part of that tube and rendered the lower section ineffective for raising steam.

This was the most economical, strong, fit-for-purpose shape that could be achieved with the material then available that would support the heavy locomotives without bending or snapping. In addition it reduced the number of potentially troublesome joints to a quarter of those in the conventional cast iron track system used up to this time. George Stephenson in his survey stipulated that the L&MR should be laid with this type of rail.

The challenge of the boiler

Engineers and millwrights working with early steam plant and engines had always worked with static equipment, in collieries or winching waggons up and down hillsides. The boilers they designed were large and heavy, as there were generally no restrictions on space or weight where they operated.

The Stephensons knew they needed a breakthrough from the design of cumbersome inefficient static boilers. With their locomotive *Lancashire Witch* in 1828 a new trend emerged. Instead of the commonplace single circular fire-tube running through the cylindrical boiler with which George's previous locomotives were all equipped, two fire-tubes of smaller equal diameter were placed side by side. In order to appreciate the great significance of this change, it is worth exploring the mechanics of boilers before this innovation.

Locomotion was a successful Newcastle-built George and Robert Stephenson machine that was launched in 1825 on the Stockton and Darlington Railway. In this locomotive the fire providing the energy to boil the water to make steam burned on a domestic-style grate inside a single circular iron tube that passed through the water in the lower portion of the steam boiler. The tube was surrounded by water when the locomotive was in use. Only the upper half of this fire-tube was exposed to the flames from the fire that passed horizontally along the boiler and up the tall chimney. The lower half of the fire-tube, and therefore the lower half of the boiler as a whole, contributed little if anything to steam production.

I have a clear memory of being asked to look after the Beamish-owned replica of *Locomotion* at a gathering of several replica early locomotives at the Bowes Railway in the 1990s. When the locomotive was in full steam, it was possible to place a hand on the metal forming the bottom of the boiler barrel. It was merely warm. But the top of the boiler was well above boiling point and too hot to touch. The Stephensons must have known about this phenomenon and looked for ways to improve the efficiency of steam production. They would have known that this temperature differential was bad for the boiler's internal stresses, as well as realising that only the upper portion of such boilers was actively generating steam.

The Stephenson design of putting two smaller fire-tubes side by side in 1828's *Lancashire Witch* allowed the heat source to be moved lower down in the boiler barrel, and the gap between the two tubes would have encouraged hot water circulation into the barrel's lower regions. Although the heating surface area was only slightly increased over *Locomotion*'s single tube the effectiveness of the two fires in two tubes increased steam generation.

One of the directors of the L&MR, Henry Booth, suggested that the Stephensons could develop the idea of using two tubes further by using many more small ones in their new design to improve steam production. Henry Booth became a partner with George Stephenson in taking the innovative concept of the *multi-tube* boiler forward. Whatever the origins of the idea, and the possible partiality and conflict of

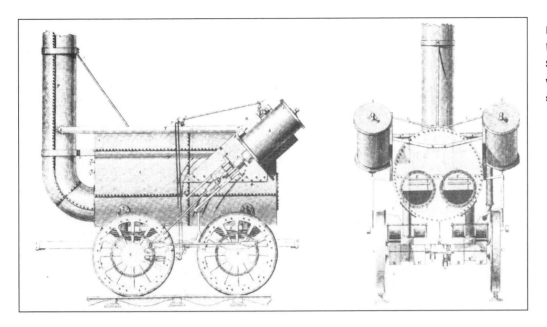

interests of Henry Booth as a Director of the L&MR, the maths makes startling reading.

Imagine the single fire-tube of *Locomotion*. If unwrapped it would form a sheet of metal 8ft long by 7ft circumference, making 56ft^2 of area. If we increase the number of smaller (3in diameter) tubes to 25, fitted in a shorter 5ft long boiler, each tube would be made from 4ft^2 of copper, giving a total of 100ft^2 in tubes alone. This obviously made a dramatic difference to the efficiency of the boiler.

Rocket was eventually designed to be built

BELOW Expanded view of boiler tubes comparing *Locomotion* and *Rocket* tube areas. *(Author)*

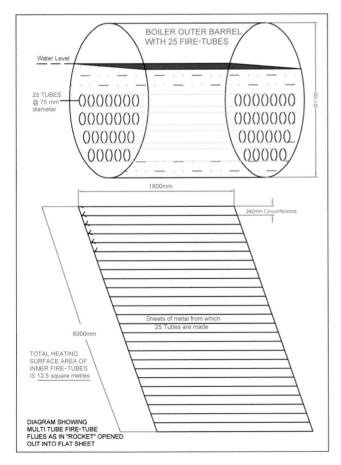

BOILER OUTER BARREL WITH 25 FIRE-TUBES

Water Level

25 TUBES @ 75 mm diameter

Ø1100

1800mm

240mm Circumference

6000mm

Sheets of metal from which 25 Tubes are made

TOTAL HEATING SURFACE AREA OF INNER FIRE-TUBES IS 12.5 square metres

DIAGRAM SHOWING MULTI TUBE FIRE-TUBE FLUES AS IN "ROCKET" OPENED OUT INTO FLAT SHEET

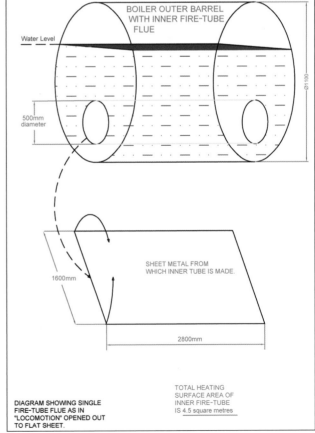

BOILER OUTER BARREL WITH INNER FIRE-TUBE FLUE

Water Level

500mm diameter

Ø1100

1600mm

SHEET METAL FROM WHICH INNER TUBE IS MADE.

2800mm

TOTAL HEATING SURFACE AREA OF INNER FIRE-TUBE IS 4.5 square metres

DIAGRAM SHOWING SINGLE FIRE-TUBE FLUE AS IN "LOCOMOTION" OPENED OUT TO FLAT SHEET.

with 25 fire-tubes leading from a separate water-surrounded firebox. The heating surface available for making steam from the flames and hot gases behind the fire-tubes was 138ft^2, as opposed to *Locomotion*'s less than 60ft^2. By more than doubling the steam-raising ability, in a much smaller, lighter boiler, the scene was set for spectacular technical improvement. The use of thinner copper tubes in place of iron also improved the heat transfer from the flames to the water inside the boiler.

Although the idea of the multi-tube boiler offered huge advantages, there was still the problem of how the fire could discharge the flames and hot gases equally down the increased number of smaller tubes. *Lancashire Witch* had two separate fires, one in each of the two tubes joined together at the base of the chimney. But if there were to be many more tubes, it would be necessary to have a separate container at the rear end of the boiler, that allowed the fire to burn yet caused the hot gases from the fire to be guided through the increased number of tubes. In addition, by surrounding the outside of that new container (the firebox) with water, even more heat could be transferred to boil the water and make more steam to improve the efficiency of the locomotive.

Steam boilers for the early locomotives generally had tall chimneys to provide a natural draught to keep the fire burning. The fire burned rather as it might in a domestic grate provided with draught from the chimney of the house. Looking at the fire one would see bright red burning coals with some smoke issuing from the chimney top. But how was the steam that had driven the pistons to be exhausted? Several of the early engineers channelled that exhaust steam into the locomotive's chimney to provide an effective silencer. One of the many objections that Stephenson and other locomotive designers received from the public and politicians was that their newfangled locomotives with their loud escaping steam would so frighten cattle that they would no longer produce milk!

The breakthrough came with what the Stephensons and others did with the exhaust steam once it was inside the chimney. By providing an upward-facing nozzle at the end of each exhaust pipe inside the chimney, a billow of steam was forced up the chimney that entrained behind it the products of combustion from the fire. This effect in turn drew fresh air into the base of the fire through the grate, which caused it to burn more vigorously. For the first time locomotive fires started to burn white hot with a corresponding increase in steam production and efficiency. The fact that this caused the coal or coke to be burned much more rapidly than before was immaterial. Coal was cheap and plentiful at this point in history.

So the Stephensons had engineered clever solutions to the two major challenges for designers of steam locomotives that hauled trains. But the directors of the L&MR were still convinced that stationary engines would be necessary. George Stephenson suggested to the directors that they hold a public competition to prove once and for all that the Stephenson locomotive type had overcome the problems of reliability and uphill locomotion with which they were so preoccupied.

BELOW Stationary engine of Richard Trevithick's design, made by Hazeldine & Co of Bridgnorth c 1803–1807, showing the exhaust steam piped along the top of the boiler into the chimney to silence the exhaust and improve the draught.

In fact father and son Stephenson had been so convinced of the bright future for steam locomotion that they had already, in 1823, set up a factory to build steam locomotives in Newcastle. It was called Robert Stephenson & Company and was to have a productive and successful future in manufacturing locomotives for many years.

The Rainhill Trials proposed

The directors of the future L&MR agreed to hold a public competition partway along the proposed route at Rainhill Levels. They invited leading engineers to create their best locomotive type to the specification that they had drawn up. They offered a prize of £500 to the designers of the winning machine.

A reading of the specification shows that the directors were looking for a lightweight, powerful, reliable machine that could sustain its performance hauling a specified load over the journey from Liverpool to Manchester and back again. By stipulating that the vehicle carrying coal and water (the tender) could be counted as part of the test load behind the locomotive they helped those engineers who were considering a conventional machine like *Locomotion* rather than those whose designs were based on a self-contained power unit carrying its own coal and water.

N° 3.

TRIAL OF THE LOCOMOTIVE ENGINES.

LIVERPOOL & MANCHESTER
RAIL WAY.

The following is the Ordeal which we have decided each Locomotive Engine shall undergo, in contending for the Premium of £500, at Rainhill.

The weight of the Locomotive Engine, with its full compliment of water in the boiler, shall be ascertained at the Weighing Machine, by eight o'clock in the morning, and the load assigned to it, shall be three times the weight thereof. The water in the boiler shall be cold, and there shall be no fuel in the fire-place. As much fuel shall be weighed, and as much water shall be measured and delivered into the Tender Carriage, as the owner of the Engine may consider sufficient for the supply of the Engine for a journey of thirty-five miles. The fire in the boiler shall then be lighted, and the quantity of fuel consumed for getting up the steam shall be determined, and the time noted.

The Tender Carriage, with the fuel and water, shall be considered to be, and taken as part of the load assigned to the engine.

Those Engines that carry their own fuel and water, shall be allowed a proportionate deduction from their load, according to the weight of the engine.

The Engine, with the Carriages attached to it, shall be run by hand up to the Starting Post, and as soon as the steam is got up to fifty pounds per square inch, the engine shall set out upon its journey.

The distance the Engine shall perform each trip, shall be one mile and three quarters each way, including one-eighth of a mile at each end for getting up the speed, and for stopping the train; by this means the engine with its load, will travel one and a half mile each way at full speed

The Engine shall make ten trips, which will be equal to a journey of thirty-five miles, thirty mi es whereof shall be performed at full speed, and the average rate of travelling shall not be less than ten miles per hour.

As soon as the Engine has performed this task, (which will be equal to the travelling from Liverpool to Manchester,) there shall be a fresh supply of fuel and water delivered to her, and as soon as she can be got ready to set out again, she shall go up to the Starting Post; and make ten trips more, which will be equal to the journey from Manchester back again to Liverpool.

The time of performing every trip shall be accurately noted, as well as the time occupied in getting ready to set out on the second journey.

Should the Engine not be enabled to take along with it sufficient fuel and water for the journey of ten trips, the time occupied in taking in a fresh supply of fuel and water, and shall be considered and taken as part of the time in performing the journey.

J. U. RASTRICK, Esq. Stourbridge, C. E.
NICHOLAS WOOD, Esq. Killingworth, C. E. } Judges.
JOHN KENNEDY, Esq. Manchester,

Liverpool, Oct. 6, 1829.

Rocket is conceived

In 1829 the Rainhill Trials proposal was the starting gun for the Stephensons together with Henry Booth to push their thinking to new levels. They looked back at the significant innovations over the recent years of steam locomotive development. Although father and son were often separated by their demanding work duties, they corresponded regularly on how their new design of locomotive might be developed. This correspondence provides fascinating insights into their innovative designs as well as their professional working relationship. Nowadays, no doubt, eyebrows would be raised about Henry Booth's role as both poacher and gamekeeper, but in those days there was no stipulation asking company directors to declare their interests.

The Stephenson design concept was to build a lightweight locomotive that would win the Rainhill competition, rather than be a permanent passenger or freight haulage machine for operating the L&MR in the long term. They understood that the method of coupling both sets of wheels together to get four-wheel drive, as with *Locomotion* and *Lancashire Witch*, caused power to be lost in the rotating side-rods. They also knew that four-wheel drive rigidly coupled meant that all four wheels had to be of exactly equal diameter to be free-running. Patrick Stirling, a great steam locomotive designer of the Great Northern Railway, was (much later) to claim that the resistance caused by coupling sets of driving wheels together was 'like a laddie running with his breeks down'! By only driving two wheels on their new machine, they were freed of the constraints of four-wheel-drive locomotives. The Stephensons limited the weight of *Rocket* to four-and-a-half tons, and – knowing that the locomotive had only to pull three times its own weight – reasoned that it only needed one pair of driven wheels.

So, between 1828 and 1829, *Rocket* was conceived and built at the Newcastle factory. It was an improved, lightweight, two-wheel-drive multi-tubed locomotive with an external firebox. Leaf springs on all four wheels were incorporated to ensure that they remained in contact with the rails at all times. This also helped to avoid overloading the track in uneven sections.

Putting the challenge on the line

Producing large, fit for purpose driving wheels at this time was a further technical challenge for early locomotive engineers. The Stephensons wanted to take full advantage of their lightweight boiler design by putting their new machine on wooden driving wheels with iron rims, rather than the clumsy composite metal ones they had used on *Locomotion*. Wooden wheel technology was well advanced and sophisticated, as nearly all vehicles had been horse-drawn up to that time. However, no one had attempted to drive the wheels round by pushing on the wooden spokes as was proposed for the wheels designed for *Rocket*. The illustrations of *Locomotion* and *Lancashire Witch* show how these problems

BELOW *Rocket* as built in 1829 and as it took part in the Rainhill Trials. Note the strengthening arm on the driving wheel to resist the turning forces from the pistons. *(John Glithero)*

Scale 12 6 0 1 2 3 4 5 6 7 8 9 10 of Feet

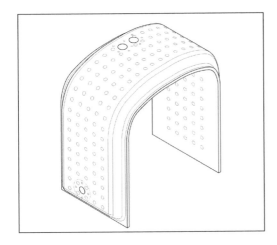

ABOVE Firebox of *Rocket*, drawn by John Glithero to take account of the twist that was in the box when it was delivered in 1829. This distortion made it necessary to modify the mountings for the boiler. *(John Glithero)*

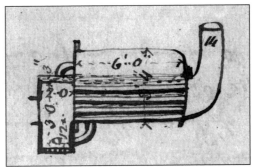

LEFT John Rastrick's sketch of *Rocket*'s boiler at the Rainhill Trials.

LEFT John Rastrick's sketch of *Rocket*'s firebox showing the stay pattern that prevented the flat sides from bulging under pressure.

were addressed in previous locomotive designs with metal wheels. Stephenson's solution for *Rocket*'s wheels was to design a wrought iron strap between the driving wheel centre and the wheel rim. This iron strap transferred the driving force from the spoke mid-point to the wheel rim and can be clearly seen in the diagram at the bottom of page 22.

The external firebox was a further technical challenge. The Stephensons designed a rectangular container with its open front attached to the tube area of the boiler. The box had an open bottom fitted with a grate on which the fire burned, supplied with air from beneath. The top and sides of the box consisted of a copper water jacket arranged with large copper water-circulating pipes that joined the firebox outer to the water in the main boiler barrel and thereby improved the amount of heating surface available. The end of the box nearest the driver was lined with firebrick to absorb the heat from the combustion, and there was a door in this end through which the fireman added new fuel. Manufacturing this double-skinned casing to enclose the fire was going to be a real challenge for the Stephensons, as its shape was unusual and awkward.

Having an external firebox meant that the boiler barrel and the firebox were two physically separated components that needed to be held tightly together. In previous Stephenson designs the components of the wheel and axle mountings had been fastened to the underside of the boiler barrel. The new design incorporated a stepped bar frame of wrought iron that was effectively the chassis of *Rocket*. The Stephensons did not directly refer to the importance of *Rocket*'s underframe but they must have realised that fastening axles to a boiler that expanded as it got hot was eventually going to land them in trouble if they did not incorporate some form of chassis into the design.

In the archive at the National Railway Museum there is a series of copied handwritten letters penned by Robert Stephenson in Newcastle to Henry Booth, keeping him informed of progress on *Rocket* in 1829. Unfortunately the other half of the correspondence has not survived, but Robert is very thorough at keeping the Liverpool team up to date with progress. He asks Henry Booth to keep his father informed of developments, as time was very tight to meet the deadline for the Rainhill Trials. Maybe George Stephenson was not confident when it came to writing and reading letters.

Significantly, Robert assumed that the hydraulic testing of the completed boiler was going to be straightforward, yet this was the very first proper multi-tubular boiler. It is likely that up to this point no one had considered the weakening effect of having 25 large holes for the copper tubes in the wrought iron end plate of the boiler barrel. Despite Robert's confidence, when first tested the end plate of the boiler pushed out under the initial hydraulic pressure and consequently weakened the fastening of the 'clunked' copper tubes (see page 59) in the tube-plate of the boiler barrel. Robert immediately ordered more longitudinal stays to be installed in the barrel.

He also wrote to Henry Booth and tactfully questioned the sense and need for the hydraulic test to be conducted at the high pressure of 150psi, as stated in the Rainhill Trials documents. He suggested that 100psi test pressure would be more appropriate, as that was still double the working pressure in steam. The rule for small steam boiler testing nowadays is to use an initial hydraulic test of twice working pressure, followed thereafter by a maximum of one-and-a-half times working pressure on subsequent tests. So Robert Stephenson's instinct for what was appropriate has been borne out subsequently by many years of locomotive boiler experience and safety testing. Henry

Booth must have concurred at least partially with the request, as in the next letter Robert Stephenson reports successful hydraulic testing of the boiler up to 120psi.

But what makes compelling reading in the letters is the level of industrial espionage that appears to be going on between the Rainhill Trials' proposed competitors. We know that the firm of Robert Stephenson & Co was contracted to supply the cast iron cylinder castings for Timothy Hackworth's *Sans Pareil*, so when the cylinders were completed he used the opportunity of delivering them to Hackworth to have a really good look at *Sans Pareil*'s boiler. These were the same cylinders that the luckless Timothy Hackworth claimed after the Rainhill Trials were one of the causes of his locomotive's poor performance. In the next letter Robert Stephenson promises Henry Booth that he will send details with a sketch of *Sans Pareil*'s boiler. He did so, stating that he thought Hackworth's design of semi-cylindrical water collar extension above the fire was very ingenious; but he believed that in spite of this ingenuity Hackworth's engine was going to be too heavy when completed. He thought that *Rocket* would be a full 13cwt lighter than *Sans Pareil*.

A most startling revelation came in a letter of 31 August to Henry Booth, when the *Premium Engine* (as *Rocket* was called before being named) had successfully withstood the 120psi boiler hydraulic test. Timothy Burstall, referred to in this letter, was one of the competing locomotive builders at the forthcoming Rainhill Trials.

Robert Stephenson wrote:

'Mr Burstall Junior from Edinburgh is in Newcastle. I have little doubt for the purpose of getting information. I was extremely mortified to find that he walked into the manufactory this morning and examined the Engine with all the coolness imagined before we discovered who he was. He has however scarcely time to take advantage of any hints he might catch during his recent visit. It would have been as well if he had not seen anything. I will write you on Wednesday Evening or Thursday Morning.
Yours faithfully
Robert Stephenson.'

BELOW Stay pattern at the front of *Rocket*'s boiler. Several of these stays were put in after the copper tubes had been inserted. This was to stop the front tube-plate bulging out and damaging the tube ends during the hydraulic test of the boiler when new.

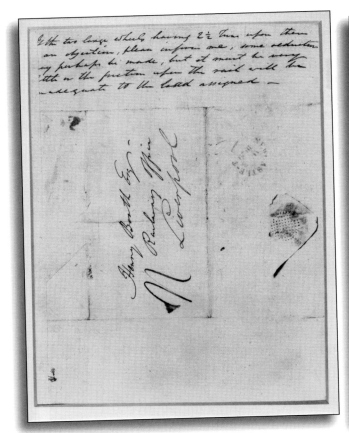

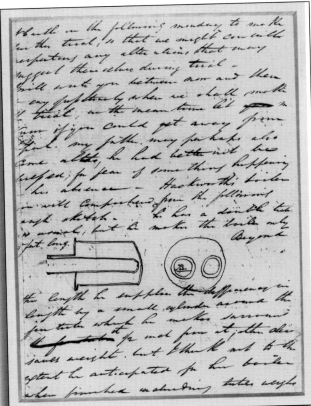

The Rainhill Trials

Given the short time available between announcing the Rainhill Trials in April 1829, and the actual Trials date set for Tuesday 6 October 1829, it is remarkable that five separate machines were ready in time to be entered for the competition. Over 10,000 curious members of the public, as well as many engineers and scientists, arrived at Rainhill Levels eager to witness this event. Public stands were erected for the crowds flocking to view the great spectacle, and a grandstand was built at the end of the marked one-and-three-quarter miles test course.

The Trials were conducted rather in the manner of a horse-race meeting, with the site at Rainhill being flat for over a mile. The locomotives were required to draw three times their own weight over the equivalent journey from Liverpool to Manchester and back again. These distances of 36 miles each way were achieved by repeated reversals over the course, with a one-eighth of a mile acceleration and deceleration section at the ends of the measured distance. The locomotives were expected to achieve at least 10mph once they were up to speed and on the measured section. Then, after a pause to take on water and more fuel, the locomotives had to perform another ten reversals to travel the equivalent of 36 miles back again.

Three independent judges were appointed by the L&MR to witness and oversee the Trials and to

ABOVE Robert Stephenson's letter to Henry Booth with covert details of *Sans Pareil*'s boiler, 21 August 1829.

BELOW An artist's impression of what was supposed to take place at the Rainhill Trials, showing the crowded public grandstand, the three principal competitors and the famous skew bridge at Rainhill.

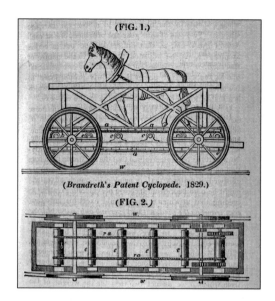

(FIG. 1.)

(Brandreth's Patent Cyclopede. 1829.)

(FIG. 2.)

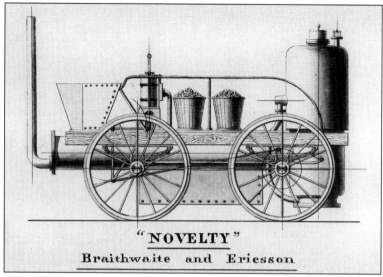

"NOVELTY"

Braithwaite and Ericsson

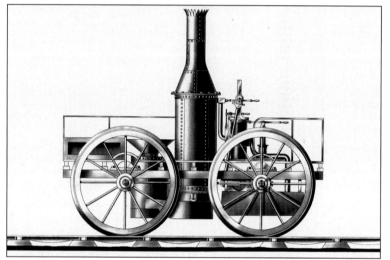

ABOVE LEFT Thomas Brandreth's *Cycloped* horse-powered entry, with gears to reverse direction of motion. Interestingly it appears to have a two-speed arrangement on the drive gears from the belt on which the horse walks.

ABOVE John Braithwaite and John Ericsson produced their so-called 'London Engine', *Novelty*.

LEFT Timothy Burstall's *Perseverance*.

BELOW LEFT George and Robert Stephenson's *Rocket*.

BELOW Timothy Hackworth's *Sans Pareil*.

The ROCKET of Mr Robt Stephenson of Newcastle.

"SANSPAREIL"

Timothy Hackworth

ensure that the rules were adhered to. They were Nicholas Wood, John Rastrick and John Kennedy. It is fortunate that John Rastrick was meticulous in his note-taking and recording during the trials, and even more fortunate that these notes and sketches survive with *Rocket*'s remains in the Science Museum Group collection.

The Trials began with demonstrations by the competitors that were ready to perform on Tuesday 6 October. Out of a total of ten original entries only five actually showed up for the contest. These were *Cycloped*, *Novelty*, *Perseverance*, *Rocket* and *Sans Pareil* (see page 26), only three of which were truly contenders for the prize: *Rocket*, *Novelty* and *Sans Pareil* (right).

Once the trials were declared officially open, *Novelty* and *Rocket* performed demonstrations for the crowds. Surprisingly *Novelty* soon became the crowd's favourite, with its apparently effortless motion and lack of moving parts on the outside. We are perhaps given a taste of the competitive climate of the event, when on one occasion *Novelty* managed to achieve 30mph free-running in front of the crowds without its test load. In response, George Stephenson – no doubt exasperated – uncoupled the tender and test carriages from behind *Rocket* and drove the lightweight locomotive past the crowd at an astonishing 35mph to re-establish his locomotive's superiority. That must have been an awe-inspiring spectacle for the crowds!

Timothy Hackworth's *Sans Pareil* was in trouble right from the start, having been declared overweight within the stipulated rules for the four-wheel design. The judges nevertheless made the decision to allow it to take part in the competition, possibly to compensate for the large number of no-shows at the start, and no doubt to please the large crowd that had assembled.

On the second day of the proper Trials it was *Rocket*'s turn to compete formally with its calculated test load of two open carriages loaded with ballast. It managed to complete the two 35-mile sections, with a midday break to take on fuel and water. *Rocket* achieved a 14mph overall average over the measured sections and a 24mph maximum. This spectacular performance was what the crowd had come for and they were not disappointed.

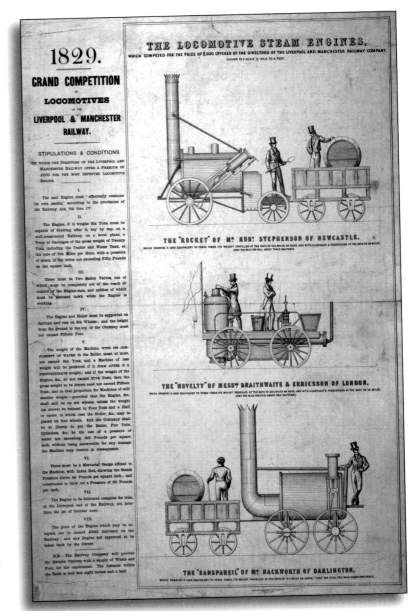

The only serious threat to *Rocket* was from *Novelty*, which like *Rocket* had an innovative boiler and, as has been said, was initially the crowd's favourite. *Novelty* set off at a cracking pace on its formal trial, but unfortunately on the third leg one of the crew turned off the feed water completely instead of diverting it back to the water tank. This caused the water feed pipe to rupture and the Trial had to be abandoned until repairs were complete. These took the rest of the day. The next day *Novelty* came out to perform again, but despite the repairs all was not well. The previous day's spell of running without a water feed after the pipe burst had caused damage within the boiler. This meant more delay. *Sans Pareil* was being frantically worked on to get it ready, but in the meantime

ABOVE The three competing contestants, *Rocket*, *Novelty* and *Sans Pareil*.

Rocket entertained the crowds by giving public rides in carriages down and up Whiston Plane to the west of the stands. George was able to demonstrate beyond doubt to everyone there that his locomotives had no need of stationary winding engines on the hills of the L&MR!

At last Timothy Hackworth's engine was also ready to compete. Hackworth was scheduled to run on the second Monday of the Trials, but being a devout Methodist he would not do the final preparation work to his locomotive on the preceding Sunday. Although *Sans Pareil* was declared 600lb overweight within the rules (Hackworth claimed that the scales were inaccurate) it was allowed to compete. However, with its single return flue boiler it was an engine of the old school of design, and disaster struck poor Hackworth halfway through the course when his water feed pump smashed, leaving *Sans Pareil* with a considerable fire and a shortage of water. The ensuing low water level and high temperature damaged the crown of the fire-tube, and Timothy Hackworth's dreams of winning the prize were over as he withdrew from the competition.

Novelty came back for another attempt but the damage previously sustained because of the boiler low water was too great for an effective on-site repair. On the fourth leg of the re-run its boiler failed spectacularly in a large cloud of steam. Braithwaite and Ericsson's dreams of winning the prize were also over.

Timothy Burstall was unlucky. His machine *Perseverance* had been damaged during unloading for the competition, and never made the stipulated 10mph. *Cycloped*, with its horses powering a treadmill, was never a serious contender.

As the other competitors failed to do fully what was required the Trials became a showcase for the Stephensons' *Rocket*, the only contender to fulfil the stipulated conditions, and it performed for the delight of the crowds. It was declared the outright winner. The £500 prize was awarded to the Stephensons and, of course, shared with Henry Booth, whose multi-tube boiler scheme enabled *Rocket* to achieve the technological equivalent of the Moon landing of its day.

Not everyone was happy with the outcome, though. There was a contemporary report in a respectable journal of the day called *Mechanics' Magazine*, which, on 31 October 1829, published 26 pages of details of the Trials, concluding:

'The Directors had no alternative [but to award the prize to the Stephensons], since "The Rocket" was the only engine which fulfilled the conditions of the competition. There are people here, however, who think that the interests of the public would have been quite as well served, had the directors adjudged the premium on a more general view of the matter, and conferred it on that engine which is, upon the whole, "the most improved." '

Robertson, the magazine's editor and the author of this piece, was a close friend of

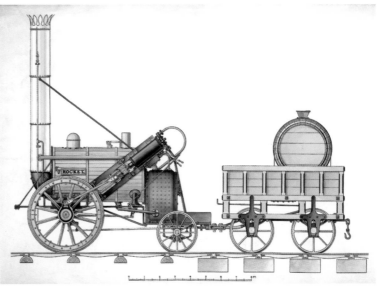

Braithwaite and favoured the so-called 'London engine', *Novelty*. Journalistic spin was clearly already alive and well in the early 19th century! He went on to write:

'…we believe we speak the opinion of nine-tenths of the engineers and scientific men now in Liverpool—that it is the principle and arrangement of this London engine which will be followed in the construction of all future locomotives.'

He could not have been more wrong. Team Stephenson had achieved much more than simply winning the Rainhill Trials prize on that October day at Rainhill. It was their machine that set the pattern for successful steam locomotive design for the next 130 years. British Railways' last steam locomotive, built in 1965, had the same principle design features as *Rocket*. Although small further developments in design occurred over the decades, none matched the great leap forward demonstrated so ably and publicly by *Rocket* at the Rainhill Trials.

Rocket after Rainhill

Once the partying was over after the Trials, there was the serious business of building a successful intercity railway to be addressed. The directors authorised the purchase of *Rocket*, and the Stephensons were rewarded not just by receiving the competition prize but also by the directors placing an order with Robert Stephenson & Co in Newcastle for several similar *Rocket*-type locomotives, to be produced in time for the opening of the L&MR later in 1830.

The experiences gained from operating *Rocket* at Rainhill and along the incomplete Liverpool to Manchester railway track taught the Stephensons that more improvements to the locomotive could be made. It was sensible to incorporate these improvements into the designs of the subsequent order for more locomotives, as well as incorporating them into *Rocket* whenever it was returned to the works in Edge Hill or Newcastle for repairs.

Rocket was involved in a number of minor mishaps and collisions during its early career. Each opportunity to repair damage to the machine gave the manufacturers the chance to update it to the latest specification. Improvements included reducing the angle of the cylinders nearer to the horizontal in relation to the wheels to reduce 'waddling' caused by the vertical component of the drive system from the steeply inclined cylinders. Also, one of the limiting factors of the original prototype *Rocket* doing a full day's work was its need to keep stopping for cleaning out the smokebox. The char and ash from the fire was carried forward towards the base of the chimney by the effect of the blast pipe, and deposited in the small container at the front of the boiler. The smokebox was therefore enlarged to match the diameter of the boiler so it was able to hold a full day's collection of char.

RIGHT Chat Moss with two trains crossing. This 12-square-mile peat bog, situated on the route of the L&MR five miles west of Manchester, was a huge challenge for George Stephenson. The present Liverpool to Manchester railway runs over George Stephenson's prepared route.

RIGHT Five-pound note showing some of George Stephenson's achievements. *(Author)*

FAR RIGHT *Rocket* box of matches.

LOCOMOTIVE ENGINE, "THE ROCKET," 1830.

RIGHT *"Northumbrian"* (not *Rocket*, as captioned in the original lithograph) arousing public interest.

In addition there was a need for coupling and buffering gear at the front of the machine, so that *Rocket* could push, as well as pull, the waggons or carriages. *Rocket* was needed in a less glamorous role than that of star performer at the trials – to move the huge amounts of material that were needed for George Stephenson's great challenge of crossing Chat Moss, where the railway was floated across that infamous bog on a bed of brushwood. Although the directors declared that Stephenson's approach to the problem of crossing the deep bog was madness, it is significant that today's modern railway between Liverpool and Manchester still crosses on the same sub-structure.

The very success of *Rocket* led to its rapid development and modification, whereas machines like *Sans Pareil* and *Novelty* remained virtually untouched. That is why members of the public are sometimes disappointed by the appearance of *Rocket* as it is presented at the Science Museum. They are expecting to see the brightly painted yellow Rainhill Trials winner they have seen on a box of matches or a five-pound note. But the Science Museum's *Rocket* is a working engine at the end of a busy life, rather than an iconic competition winner.

The remains of *Rocket* in the Science Museum are perhaps more easily explained by looking at the lower illustration opposite, showing the development of the *Rocket* design in the second batch of engines – in this instance *Northumbrian*, with different firebox, smokebox and near horizontal cylinders.

It is interesting to speculate at what stage a great technological breakthrough like *Rocket* becomes an icon and its status changes to the point where it becomes an object of curiosity to be placed in a museum. Parallels in other fields are the Moon landing in the 1960s, where only the returning capsule survived to be displayed. The *Golden Hind*, Sir Francis Drake's ship used in the circumnavigation of the world, was displayed at Deptford upon its return for nearly 100 years. Harrison's timepiece that solved the longitude problem is now displayed at Greenwich. The Wright Brothers' first successful aircraft is in the Smithsonian Institute in the USA. The Norwegian *Kon Tiki* raft, that showed in 1947 that Polynesian Islanders could have come from South America, is displayed in Oslo. All these iconic objects were mothballed shortly after their significant achievements.

However, *Rocket* was put to work commercially and was developed into an improved version of its original self as early steam locomotive technology moved forward rapidly. *Rocket* was a working construction locomotive hauling loads of spoil and materials to build the remainder of the railway towards Manchester. In addition *Rocket* was used by the directors to entertain interested parties wishing to have their first taste of the new steam-locomotive-powered railway experience, running a sort of 'trips round the bay' exercise between Liverpool and Rainhill. The purpose of this early PR exercise was to spread the word that here was a new and exciting method of travel that the general public could embrace and enjoy.

These trips will be discussed in more detail later in the book. A quiet revolution was taking place – *Rocket* had showed that it could easily climb the steep hill up to the Rainhill Levels with a full load of passengers from Liverpool Crown Street. This was the steepest gradient on the line, and the success of *Rocket* put paid to any ideas the directors of the L&MR might have held in support of using stationary steam winding engines to haul their trains up and down the inclines. Without the delays inevitable in coupling and uncoupling trains on to haulage ropes, the prospect of completing the journey from Liverpool to Manchester in less than three hours was becoming a reality!

However, even with the updating of the locomotive when modification opportunities arose, *Rocket* was never going to be able to compete with the heavier, more powerful improved machines that were coming out of the Newcastle Works of Robert Stephenson & Co. *Rocket* had been built for speed with light loads, and it gradually became less frequently used and was often to be found in storage.

But there was to be one further occasion when *Rocket* again took centre stage, though this time for all the wrong reasons. That was the grand opening of the completed L&MR on 15 September 1830.

Opening

OF

THE LIVERPOOL AND MANCHESTER

RAILWAY,

WEDNESDAY, 15TH SEPTEMBER, 1830.

CHAS. LAWRENCE, CHAIRMAN.

THE BEARER OF THIS TICKET IS ENTITLED TO SEAT No. · 34 ·
NORTH STAR'S TRAIN.

YELLOW FLAG.

ENTD *I. Read*

ABOVE Opening day ticket for the Liverpool and Manchester Railway.

The opening of the Liverpool and Manchester Railway

The opening of the L&MR was to be a showcase occasion to the rest of Britain. The event was planned as a railway procession from Liverpool to Manchester and back. The Duke of Wellington, Prime Minister at that time, was guest of honour and a special wide carriage was built to carry the duke and his entourage on his journey to Manchester and back. Eight special trains were to take part in a cavalcade with 700 special guests. In order to maximise the effect of the cavalcade it was decided to use both tracks concurrently in one direction, with all crossovers and points locked out of use. The duke's special train for his entourage, with its palatial extra-wide coach, was to be hauled by one of the new state-of-the-art Stephenson locomotives from Newcastle, *Northumbrian* (see lower illustration on page 30), while the other seven trains would be on the adjacent track with their own carriages so that the duke was highly visible to both the travelling and spectating public. The plan was for the trains hauled by *Rocket* and several of its newer sister locomotives to be re-marshalled to return from Manchester after suitable celebrations.

All went well in the early stages of the pageant, and the cavalcade stopped as arranged to replenish the locomotives' water tenders at Parkside, which was just over halfway along the route. This was seen as an opportunity by one of the guests, local MP William Huskisson, to get down and cross the tracks to talk to the duke, who he was anxious to impress on such a prestigious occasion. As *Rocket*'s driver, White, under the direction of Joseph Locke, also on the footplate of *Rocket*, approached at the head of the train on the track adjacent to the duke's special carriage – stationary by this time – he could see there was little clearance between *Rocket* and the extra-wide carriage. But *Rocket* was unable to stop in the short time available (of which more later). Huskisson was holding the open door of the duke's carriage when, desperately, he realised that he was going to be trapped between the oncoming locomotive and the carriage. He fell backwards into the path of the oncoming train as the door of the carriage was wrenched off, and the locomotive and train ran over his upper leg, seriously injuring him.

This incident was a public relations disaster for everyone, especially during such a public spectacle. In order to try and save Huskisson's life, George Stephenson uncoupled *Northumbrian* from the duke's train and, with the injured MP Huskisson laid on a flat waggon that had been used as a mobile podium for the military band, sped forward in a high-speed dash towards Manchester to get medical assistance at Eccles. Unfortunately Huskisson's life could not be saved.

His untimely and very public demise brought about much discussion as to the manner in which he had been run over. To us it is self-evident that *Rocket*, travelling forward at 8mph with a loaded train, could not have stopped instantly, nor taken action to avoid the accident. At the time, however, people seem to have somehow expected a train to be able to avoid a man, as a horse-drawn road vehicle might have done. However, the railway with its pre-determined route and poor wood-to-metal braking characteristics, was newfangled and strange to the onlookers. Until then brakes had been a low-priority luxury, as stopping trains was not as important as making them go.

With the benefit of hindsight it is possible to see that most railway accidents bring some sort of regulated improvement in their wake, and brakes started to be improved from this moment on. But this was too late for poor William

ABOVE Liverpool and
Manchester Railway
opening-day scene.
The trains are ready
to set off through
the Moorish Arch for
Manchester. Note the
Duke of Wellington's
extra-wide carriage,
which was to
be the undoing of
Mr Huskisson later in
the day.

LEFT Parkside station
water stop in 1831,
where the opening-day
procession stopped to
water the locomotives.

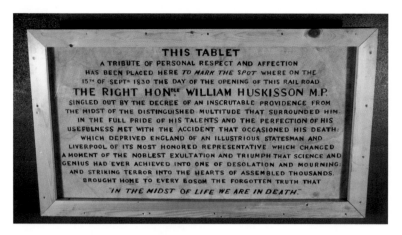

THIS TABLET
A TRIBUTE OF PERSONAL RESPECT AND AFFECTION
HAS BEEN PLACED HERE *TO MARK THE SPOT* WHERE ON THE
15TH OF SEPTR 1830 THE DAY OF THE OPENING OF THIS RAIL ROAD
THE RIGHT HONBLE WILLIAM HUSKISSON M.P.
SINGLED OUT BY THE DECREE OF AN INSCRUTABLE PROVIDENCE FROM
THE MIDST OF THE DISTINGUISHED MULTITUDE THAT SURROUNDED HIM,
IN THE FULL PRIDE OF HIS TALENTS AND THE PERFECTION OF HIS
USEFULNESS MET WITH THE ACCIDENT THAT OCCASIONED HIS DEATH:
WHICH DEPRIVED ENGLAND OF AN ILLUSTRIOUS STATESMAN AND
LIVERPOOL OF ITS MOST HONORED REPRESENTATIVE WHICH CHANGED
A MOMENT OF THE NOBLEST EXULTATION AND TRIUMPH THAT SCIENCE AND
GENIUS HAD EVER ACHIEVED INTO ONE OF DESOLATION AND MOURNING:
AND STRIKING TERROR INTO THE HEARTS OF ASSEMBLED THOUSANDS.
BROUGHT HOME TO EVERY BOSOM THE FORGOTTEN TRUTH THAT
"IN THE MIDST OF LIFE WE ARE IN DEATH."

ABOVE The Huskisson Memorial, which was vandalised and then removed to the National Railway Museum for repair and safekeeping. A replica was installed at Parkside.

Huskisson. A replica memorial stands to this day on the L&MR at the spot at Parkside where he was run down by *Rocket*. The original damaged and restored plaque is in the National Railway Museum's collection.

Rocket after the Liverpool and Manchester Railway opening

Just as innovative computer systems were rapidly made obsolete in the early days of information technology, so the developing technology of the early steam locomotive was fast moving and ever changing. *Rocket* and the locomotives developed by Robert Stephenson & Co from the prototype were not ideal for pulling the increasingly heavy trains that the new and successful intercity railway needed. Therefore the company developed a new type of locomotive called the *Planet*, where the cylinders were turned round and moved inwards to be encased in the base of the smokebox and the drive was arranged to the large pair of wheels at the rear of the locomotive. The *Planet* design brought the drive mechanism between the wheels inside the frames, necessitating manufacture of a crankshaft (as would be found on a pedal car). *Planet* was another great technical leap forward, and *Rocket* was frequently sidelined and stored out of use.

In 1832 a branch line to Wigan was built off the L&MR at Parkside. *Rocket* was used to operate this new branch for the initial period of its operation, but a head-to-head collision with a freight train put it back in the Newcastle Works for repair, and updating of its running gear. *Rocket*'s periods of storage meant that it was available for conducting experiments by various parties full of good ideas about how to improve the concept of a steam locomotive. One significant idea in 1834, a bold and innovative move, was by private inventor Thomas Cochrane, Lord Dundonald. He wanted to create a purely rotary driving mechanism that surrounded the front axle to eliminate conventional pistons, cylinders and cranks. A vane was driven round inside the cylinder by the steam. In many ways, it was like the Wankel rotary internal combustion engine that failed to break the near monopoly of the multi-cylinder inline petrol engine in the 1990s. Lord Dundonald's innovation, which was incorporated into the front axle of *Rocket*, was a failure because a satisfactory seal could not be obtained between the rotating vane and the stationary cylinder, and *Rocket* went back to being a spare conventional locomotive at Edge Hill. The rotary petrol engine idea revived by Wankel in the 20th century suffered from rotor sealing problems in the same way that Lord Dundonald's innovative engine did with *Rocket*.

A new opportunity came for *Rocket* in 1836, when the Earl of Carlisle's colliery agent, James Thompson, was searching for a lightweight

BELOW *Planet* locomotive of December 1830, from Robert Stephenson & Co. Designed for the L&MR, it built on the success of *Rocket* and incorporated many significant improvements. The cylinders were contained under the smokebox, which necessitated the use of a cranked axle on the driving wheels.

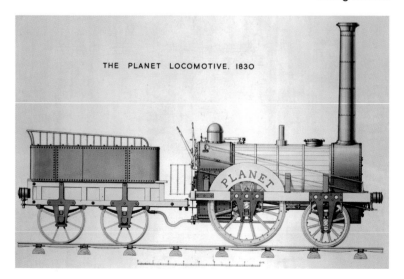

THE PLANET LOCOMOTIVE. 1830

PLANET

steam locomotive to pull coal waggons on their lightly constructed colliery railway system near Naworth in Cumberland. Thompson bought *Rocket* and its tender from the L&MR board for £300, and it was shipped north to Carlisle. *Rocket* appears to have had little use in this proposed new role but was regarded with great affection by James Thompson, who kept it nonetheless. Dr Michael Bailey quotes a local newspaper article:

'It now stands no less a monument to the genius of the inventor than as a mark of esteem in which his memory is held by Mr Thompson who has to boast of an unvarying and unbroken friendship with Mr Stephenson of nearly 25 years standing.'

This was the first recorded sign of a desire to keep and cherish *Rocket* as an icon and prevent it from being scrapped as it became less capable of keeping up with modern locomotive trends. James Thompson kept *Rocket* stored at Kirkhouse, but he brought it out for a publicity stunt in the 1837 General Election, speeding with the ballot boxes from the Alston constituency over part of the 48km journey to Carlisle.

Rocket as a museum exhibit

Plans for the 1851 Great Exhibition initially included the possibility of exhibiting *Rocket* alongside current locomotives of the period, so displaying two decades of steam locomotive development to the public. The engine was returned to the Newcastle Works of Robert Stephenson & Co with a view to readying the engine for display. The period of storage, however, had taken its toll on *Rocket* and many non-ferrous components were missing, believed to have been stolen for their worth as scrap metal. The job of replacing so many missing parts was too taxing for Robert Stephenson & Co, and *Rocket* remained in storage at the Newcastle factory where it had been built.

By 1862 the Patent Office in London was starting to build a collection of significant machines from the industrial revolution, to which James Thompson's widow donated *Rocket*. Robert Stephenson & Co were then faced with

the dilemma that faces many museum curators when presented with an object that had been extensively modified throughout its life – that is, whether to display the object as found or to try and restore it to its fame and glory days at Rainhill: the latter decision would have involved discarding some original *Rocket* material and substituting replica components, which could be interpreted as unauthentic. The dilemma was compounded by the lack of knowledge about what the machine had actually looked like at Rainhill. There were no contemporary

BELOW Front page of *The Graphic* newspaper, showing *Rocket* in 1880, in the form in which it was gifted to the Patent Office Museum. This museum became part of the South Kensington Museum in 1884, which by 1909 had split into two – The Science Museum and the Victoria and Albert Museum.

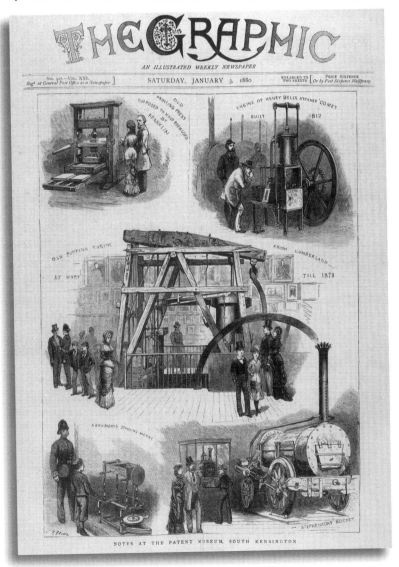

photographs, of course, but Robert Stephenson & Co did their best by producing a conjectural drawing of what they thought *Rocket* would have looked like if complete and in its final operating state. Inevitably, a compromise was reached and the final version of *Rocket*'s much modified structure was donated to the Patent Office Museum's collection.

The Science Museum curators realised over time that some of the previously made additions to *Rocket* – such as the chimney and the exhaust pipes – were conjectural, and over the next 130 years several modifications were made to correct some of the more obvious errors.

Rocket made at least three more forays out of its home in South Kensington. Firstly, in September 1941, during the Second World War, it was evacuated to safety at Brocket Hall near Welwyn Garden City, returning in June 1945. Secondly, it was loaned to Merseyside County Museums as part of the 150-year anniversary celebrations of the Liverpool and Manchester Railway between April and November 1980. It was then returned to the Land Transport Gallery in the Science Museum. Lastly *Rocket* was sent on loan to the 'Treasures of the Science Museum' touring exhibition in several cities of Japan between March and September 1998.

It was then that the National Railway Museum had the opportunity to work on a detailed archaeological investigation of the remains of *Rocket*. Unlike conventional archaeology, which digs down through the layers, recording and removing the evidence it finds, this exploration or 'dig' was only permitted on the basis of every action being totally reversible, with no original material being destroyed during dismantling and reassembly. The archaeological exploration was conducted by internationally acknowledged early steam locomotive historians Dr Michael Bailey and Dr John Glithero with the help of Peter Davidson. The detailed findings of that survey discovered much that was previously unknown, and were published in National Railway Museum's

BELOW *Rocket* in Patent Office Museum form, photographed outside South Kensington Museum c 1876, it shows several conjectural features added by Robert Stephenson & Co.

LEFT So where's the firebox? Dr Michael Bailey and John Glithero begin their exploration of *Rocket* at the National Railway Museum.

BELOW *Rocket* as displayed in the 'Making the Modern World' exhibition at the Science Museum.

definitive 2000 publication *The Engineering and History of 'Rocket': A Survey Report*.

Rocket had been changed radically during its few years of operation on the L&MR. Evidence of those changes was revealed as the locomotive was stripped and reassembled. The physical dismantling and recording was carried out in full view of the public in 1999 so that museum visitors were able to share in the excitement of the discoveries and findings as they emerged. A blackboard was used to write up the day's activities as well as to announce the findings of the day. A 'Rocket Round-up' took place on a regular basis at noon so that the public could hear a talk about the discoveries whilst the experts were at work.

Unlike a normal archaeological dig where the actual material is painstakingly removed and examined to reveal hidden secrets underneath, almost in the manner of a post-mortem, the *Rocket* 'dig' was carried out in such a way that material was removed in a fully reversible manner. This meant that when the examination was completed and the evidence gathered, recorded and investigated, all the removed components could then be carefully replaced leaving no signs of intervention and with all the evidence intact. In a way this was unusual for industrial

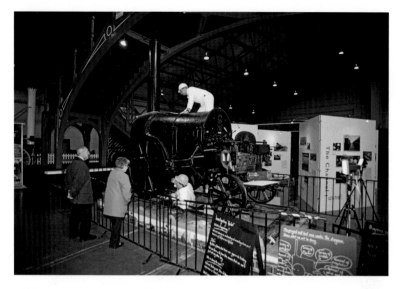

ABOVE *Rocket* 'dig' under way, with blackboard to inform the visiting public of progress.

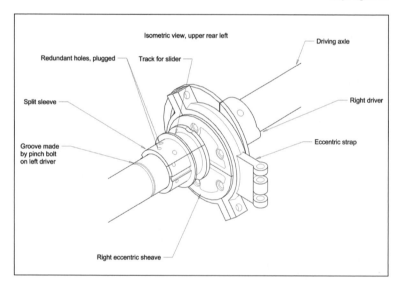

Isometric view, upper rear left

Redundant holes, plugged

Track for slider

Driving axle

Split sleeve

Right driver

Groove made by pinch bolt on left driver

Eccentric strap

Right eccentric sheave

ABOVE Drawing of assembled split sleeve that caused a sensational outcome for the author. *(John Glithero)*

RIGHT Split sleeve with keyway from *Rocket*'s front axle, discussed in the text.

archaeology, because the whole essence of a 'dig' is to work through the various layers, removing finds and recording them for posterity.

The actual work on the museum floor was accompanied by painstaking research of the Stephenson & Co ledgers and archives as well as the Science Museum and National Railway Museum files on *Rocket* and other relevant sources. So, for example, when damage to the frame in a certain place was found and recorded, it was possible to relate this to records of accident repairs carried out at both Edge Hill workshops and the Newcastle factory when locomotives were returned for repairs or upgrading to the latest specification.

For me personally one of those 'Rocket Round-up' sessions was especially memorable. It was on the occasion when the front axle and reversing eccentric cluster was being dismantled and something very special took place. (See left and below left.)

It is hard to communicate in words the excitement I felt as I handled the latest component that had just been removed, and why it resulted in one of the highlights of my heritage-engineering career. To judge that excitement in context it is necessary to fill in some technical details about mechanisms, keyways, shafts and how components slide and work together. If an engineer wants to ensure a sound sliding fit on a drive mechanism that is mounted on a shaft it is necessary to use a key. This key fits into a slot in the fixed component that engages in a rectangular slot in the sliding component, called a 'keyway'. But the slot shown in the diagram on the left needs to finish at the shoulder. So in this instance the manufacture of the key and the slot needed to be controlled precisely, to ensure that everything fitted correctly without shake or slack.

Nowadays each of the two keyways in this kind of component would be created on a vertical milling machine (see top of page 39), which would produce a round-ended slot as the cutter approaches the shoulder at the end of the slot. However, we can see that the slots on *Rocket*'s eccentric mounting sleeve (left) are square-ended. How can this be?

There is only one way that the square slot can have been produced at the Newcastle factory; that is by cutting with a hammer and chisel by hand. This is an extremely skilful process and must have been carried out by a millwright or fitter under supervision by Robert Stephenson or his right-hand man. It would have involved careful marking out of the component followed by judicious use of a hammer and sharp chisel to chip out the several cubic centimetres of material from the place where the slot was required. Careful inspection of the end of the slot showed traces of the chisel marks where the metal had been pared away.

As I held *Rocket*'s original component in my hand, I found it inspiring to see the evidence of mechanical work done so long ago. Although it does not have the cachet of, say, handling a piece of lunar rock, it does have that magical ability to connect what we curators and historians at the museum were doing with *Rocket*, to what those craftsmen had done in Newcastle with the original machine. To be able to tell this story to the museum's visitors who had gathered for the 'Rocket Round-up' that day in 1999, and to then pass the component around amongst them, getting them to put their fingers into the end of the slot and to feel and see those chisel marks, was a rare privilege. I hope some of the museum's visitors felt the same sense of connection to *Rocket*'s builders as I did.

As the National Railway Museum's mechanical engineer this magic moment ranks alongside being handed the set of drawing instruments (thought to be a gift from his father Marc Brunel) belonging to Isambard Kingdom Brunel that he had used for all his drawings (see right) – I had been asked to write a short definition of the function of each of them. The drawing instruments in their case are destined to go on display in a new 'Being Brunel' exhibition at the SS *Great Britain* Museum in the Great Western Dockyard in Bristol.

The key findings of the 1999 examination of *Rocket* are summarised in Appendix I of this manual, and changed the perceived wisdom about *Rocket*'s history from that moment on. This made the locomotive's return to take its place in the centre of the 'Making the Modern World' exhibition in the Science Museum at South Kensington in 2000 even more special.

LEFT Milling a round-ended slot using modern methods on a vertical milling machine. *(Author)*

BELOW Isambard Kingdom Brunel's drawing instruments. *(SS Great Britain)*

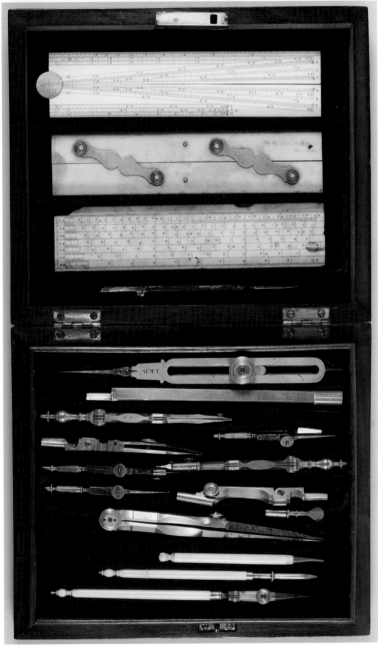

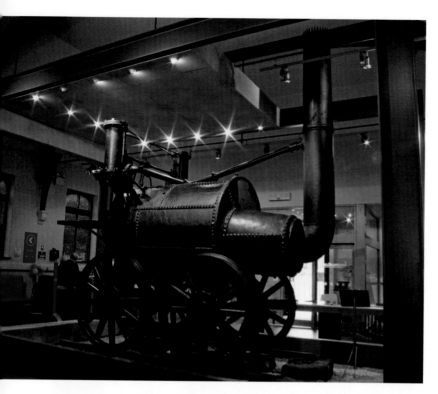

ABOVE *Sans Pareil* on display at Locomotion, Shildon.

BELOW *Novelty* in conjectural image.

What happened to the also-rans after the Rainhill Trials?

*R*ocket's principal competitors at the Rainhill Trials, *Sans Pareil* and *Novelty*, did not disappear entirely without trace from the *Rocket* story, in spite of their difficulties at the trial itself. Both locomotives were fully repaired and were bought by the L&MR Company for further evaluation. They worked on the railway in lesser roles than *Rocket*. Both machines represented different lines of development that were eclipsed by the slick, compact design of *Rocket*. *Sans Pareil*'s cylinders were enlarged to make it more powerful and it worked until it was retired from the Bolton and Leigh Railway where it had run until 1844. Then it was converted to drive a static pumping engine at Coppull Colliery, Chorley. It was from that role that it was rescued in 1863 and donated via the Patent Office collection to the Science Museum, where it joined its old rival, *Rocket*. At the time of writing, it is on display at Locomotion, the National Railway Museum's sister museum in Shildon, County Durham, as part of the National Collection. It is pleasing that it is displayed so close to where it was originally built.

Novelty fared slightly less successfully, although the directors of the L&MR did order two similar but larger locomotives working on the same principle. The original repaired *Novelty* was transferred to work for a few years on the St Helens and Runcorn Gap Railway. In 1833 it was rebuilt with new wheels and new cylinders, but it was never going to be a world-beater like its rival. In 1929 the discarded wheels and one of the two cylinders were incorporated into the full-scale non-working replica that is now displayed at the Manchester Museum of Science and Industry. The surviving cylinder, which is part of the National Railway Museum's collection, was put on display at Rainhill Community Library, where visitors can still see it.

Although *Novelty* and *Sans Pareil* were never going to match the performance and charisma of Stephenson's *Rocket* following the Rainhill Trials, we must not write these

machines off as insignificant. It seems that
Braithwaite and Ericsson only had seven
weeks between hearing of the trials at Rainhill
and having their very first railway locomotive
ready for testing at Rainhill. That alone is a
remarkable engineering achievement.

But we have not heard the end of
competition for *Rocket* under the Rainhill Trials'
'Ordeal' conditions. Chapter six describes how
the controversy was to blow up again between
three replicas of the original Rainhill Trials
contestants under conditions that were every
bit as exciting as those that pertained 170
years earlier.

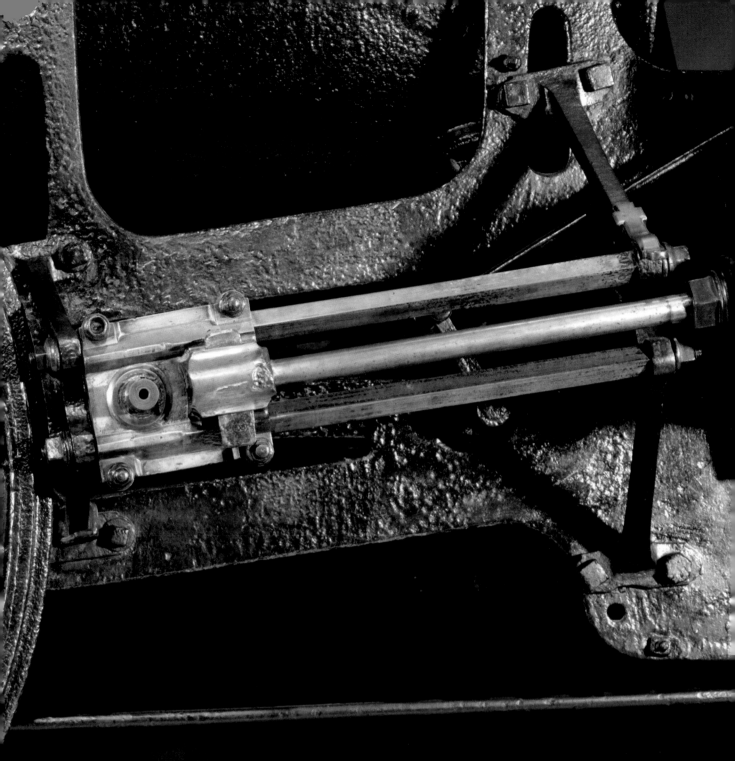

CHAPTER TWO

The anatomy of *Rocket*

Introduction: *Rocket*'s vital statistics

Table 1: *Rocket* in working order, Rainhill Trials condition in 1829	
Weight of *Rocket* locomotive	4.25 tons (4,318kg)
Weight carried on driving wheels	2.5 tons (2,540kg)
Driving wheel diameter	56in (1,422mm)
Carrying wheel diameter	30in (762mm)
Wheelbase (between axle centres)	86in (2,184mm)
Diameter of wrought iron boiler barrel	40in (1,016mm)
Length of boiler barrel excluding firebox	72in (1,830mm)
Grate area within firebox	6ft^2 (0.56m^2)
Number and diameter of copper tubes	25 x 3in (150mm)
Total heating surface within boiler	138ft^2 (12.8m^2)
Maximum steam pressure	50psi (3.4bar)
Safety valves on boiler	1 Hackworth spring, 1 deadweight
Axle suspension	4 leaf springs
Cylinder diameter	8in (203mm)
Stroke of pistons	16.5in (419mm)
Slide valve operation	Eccentric on main axle
Valve type	Flat slide valve with exhaust cavity
Reversing type	Slip eccentric with manual override
Overall length of locomotive	14ft (4.2m)
Overall height to chimney top	16ft (4.9m)

The basic design

Apart from brakes, *Rocket* in its developed form embodied all the elements that you would find in a so-called modern steam locomotive. Although there appears to be a great difference in appearance between the early and final versions of *Rocket*, the anatomy is little changed except for differences in the smokebox, the dome and the cylinder arrangement.

A multi-tubular boiler was fired by burning coke on a cast iron grate within a water-surrounded firebox external to the circular boiler barrel. Air for combustion was drawn in underneath the grate within the firebox. Steam was supplied from the boiling water under pressure inside the boiler, via a regulator valve under the control of the driver to steam chests attached to each of two cylinders. Within each steam chest was a reciprocating flat slide valve

BELOW General arrangement of *Rocket* in its original working form (left) and final form (right). *(John Glithero)*

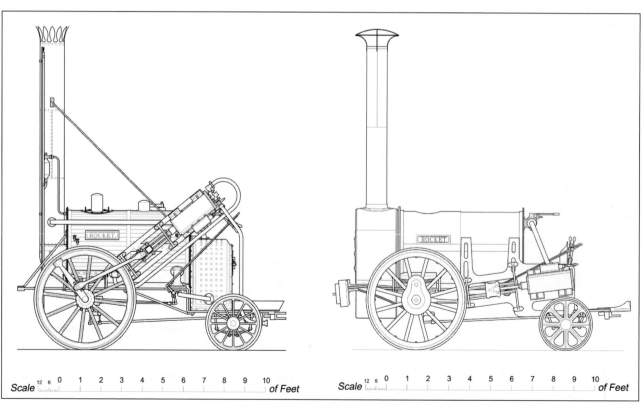

Scale 12 6 0 1 2 3 4 5 6 7 8 9 10 *of Feet*

Scale 12 6 0 1 2 3 4 5 6 7 8 9 10 *of Feet*

that distributed steam to the alternate ends of the power cylinder containing a piston. The piston was sealed into the bore by two split rings arranged in grooves around the circumference. A piston rod protruded from the lower end of the cylinder and was anchored into a guided crosshead.

The moving crosshead drove the crank pin on each driving wheel, which was pushed and pulled around by a stout connecting rod in the same way that a pedal car is operated by a child. The crank pins on the wheels were arranged to be 90° out of phase so that the locomotive always had one of its power cylinders available to deliver the maximum starting push or pull, even when the other one was at the end of its stroke and therefore ineffective.

Once the pressurised steam had pushed the piston to the end of the cylinder, the moving valves (driven from the front axle) exhausted the steam through passages that led it to the base of the chimney at the front of the boiler. The exhaust steam was directed through a nozzle to expand at speed up the chimney and in so doing entrain the gases in the area of the chimney. That in turn led to the exhausting hot gases drawing extra fresh air

LEFT Regulator of replica *Rocket* showing the steam pipes feeding to the left and right-hand cylinders. *(Author)*

LEFT Sectioned replica cylinder and valve assembly. *(Author)*

LEFT The big end of the connecting rod on the replica. *(Author)*

FAR LEFT Piston rod and crosshead assembly on replica.

LEFT Connecting rod on replica, which joins the reciprocating piston and rod to the rotating crank on the wheel. *(Author)*

ABOVE Exhaust pipe on sectioned replica, running from the cylinder to the chimney above the nameplate. (Author)

RIGHT Twin blast pipes on replica showing the narrowing of the exhaust pipe diameter through the upturned nozzle. (Author)

in through the bottom of the grate, making the fire burn ever brighter.

What is fascinating about the anatomy of *Rocket* is that if George or Robert Stephenson were confronted with the last steam locomotive built in the UK in the 1960s, they would no

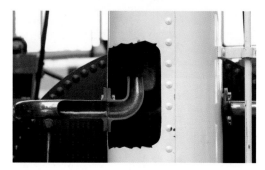

doubt be able to understand the machine and even drive it. How amazed they would have been to realise that the basic design formula they got so right in 1829 had persisted virtually unchanged for 130 years!

Why was *Rocket* so exceptional that it remained the basis for the anatomy of all modern steam locomotives? The design of *Rocket* was different from the locomotives that preceded it in that the wheels of previous engines were often fastened to the underside of the boiler itself. *Rocket*'s boiler and firebox were separate items that were joined together, and these two main boiler components were mounted on a frame.

The Stephensons wanted to get away from fastening their locomotive wheels and axles to the underside of the boiler, which caused the axles to have variable centres and stressed the boiler. A frame provided mounting points for the four leaf springs that allowed each of the wheels to cope with track irregularities. Also, the Stephensons were aware that the joint between the boiler barrel and the quite flimsy external copper firebox represented a significant structural weakness if the boiler assembly was going to be used as a structural mounting for the wheels. The iron frame in their design created a sound working chassis for the wheels and axles, which were attached to the frame by leaf springs suspended from pivoting hangers.

One exceptional feature of *Rocket* at the time of the Rainhill Trials was its tall chimney, with

RIGHT Isometric frame drawing, showing how the boiler is supported by the frame in four places. (John Glithero)

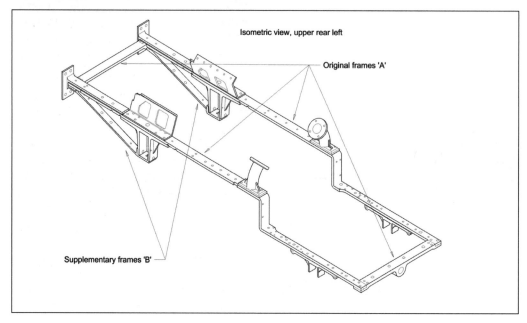

Isometric view, upper rear left

Original frames 'A'

Supplementary frames 'B'

its petal-shaped top 16ft above rail level. The chimney was tall for two reasons. Firstly, it was there to create a good natural draught for the fire when the locomotive was stationary. Secondly, the steam pressure in the boiler was measured by a U-tube manometer containing mercury, which involved the pressure in the boiler pushing against the column of mercury (the way that physicists measured pressure). The boiler pressure was set to 50psi. This would raise a column of mercury to a height of about 102in, so a tall chimney was essential to accommodate a manometer capable of displaying the boiler pressure to this level. A wooden float on top of the mercury showed the height to which the liquid had been pushed and therefore indicated the pressure to the crew. As boiler pressures started to rise several times higher than that which *Rocket* used, it was a bonus that the Bourdon tube-type pressure gauge was invented around this time. Using the mercurial U-tube manometer would have resulted in even taller chimneys, and getting through tunnels might have been a serious problem! (See illustration above right for explanation)

All railway locomotives are defined by weight, size and pulling power. In order to restrict track damage, the directors of the L&MR wrote the brief for applicants of the Rainhill Trials in 1829 with a limit on the weight of their machines. Weight restrictions continued to apply throughout the history of steam locomotives, as track and bridge engineers were always concerned about the effect and maintenance costs of heavy locomotives on their structures and tracks.

Tractive effort

Tractive effort is the force that a locomotive's driving wheels exert on the rails to pull itself and its train along the track. The friction of a metal driving wheel forcing a locomotive along on a dry metal rail generally delivers a force of about a tenth of the weight on the driving wheel without it slipping. So with an overall weight of 4.25 tons on the four wheels of the locomotive, two-thirds of the weight is likely to be carried on the larger set of driving wheels. That leaves just over 2.5 tons adhesive weight available for traction. This would give us in modern terms a tractive effort of 0.135 tons from each driving

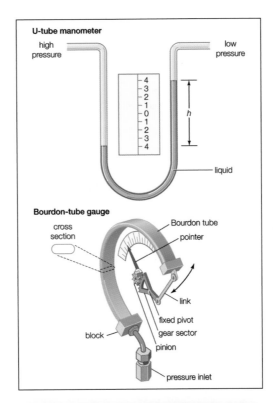

LEFT Explanation of manometer and Bourdon gauge as different ways of measuring steam pressure.

LEFT Mercurial manometer attached to chimney of the replica. A wooden float on top of the mercury indicated the pressure to the crew. (*Author*)

wheel. The driving wheels are fixed 90° out of phase and each delivers its maximum tractive force when the other is changing piston direction and delivering no force at all; so, as with a pedal car, the driving force from each piston is not truly additive but is delivered in alternating cycles.

The tractive effort of *Rocket* was considered suitable for pulling a train weight of about 15 tons at the time on level track, although the quality, efficiency and effectiveness of the bearings that the wheels and axles of the train used was nothing like we are used to today. Modern oils and lubricants did not become available until the end of the 19th century. Animal fats were used as lubrication in the bearings to achieve low rolling friction, but the ability to climb modest hills with a load behind the locomotive was an unexpected bonus of early locomotive experiments such as the Rainhill Trials. Before then the locomotive engineers had assumed that gradients would be climbed with stationary engines rope-hauling the trains up the hills, as had been the case with steam locomotive technology up to this point. This belief persisted surprisingly far into modern times, when terminal stations like London Euston, Liverpool Lime Street and Glasgow Queen Street stations were all conceived with trains being rope-hauled up the steep inclines out of those termini by stationary steam engines, with the example at Glasgow Queen Street persisting until the early 20th century. Nowadays a passenger embarking on a journey from one of these stations is barely aware that the train is climbing! Spare a thought for the brave firemen who, to slacken

the rope at the summit, walked to the front of the locomotive along the running board whilst the driver accelerated the train. The fireman then bent down and lifted the shackle off the locomotive's front buffer beam hook and dropped the moving rope into the gap between the rails as the train sped on its way.

The frames

Those of us familiar with the design of UK steam locomotives are used to seeing the frames as two parallel deep plates with the axles passing at right angles through gaps in the plates. This allows limited up-and-down movement of the wheels outside the plates, controlled by springs. The boiler and the rest of the locomotive are attached to the frames as a rigid self-supporting unit. In contrast, *Rocket*'s frame requirements were to carry the two sections of the boiler and support them as though they were a rigid unit. Robert Stephenson's design had two stepped longitudinal flat bars of wrought iron bent into a 'Z' shape. These provided an operating platform at the rear of the locomotive and supported the two separate parts of the boiler and the cylinders. The wheels and axles were attached with springs to the frame at the front and back.

This configuration was more akin to the practice we would see later in the USA as steam locomotives developed, where 'bar frames' became the norm for steam locomotive construction as opposed to the 'plate frames', described above, favoured in UK designs. *Rocket*'s boiler was fastened firmly to the frames by the brackets and the trunnions (see bottom of page 46 and below). There must still have been some minor distorting effect from boiler expansion between hot and cold. As the design of steam locomotives progressed, expansion over the length of long boilers became an increasing technical challenge. The solution was to anchor the front end of the boiler rigidly to the frames and let the back end slide on captive expansion joints. The system was similar to that which we might find in a modern metal bridge, where one part is firmly anchored and the length is free to expand and contract on slides.

As the stresses in a pressure vessel like a boiler became better understood, new designs ensured that none of the stresses that should be carried by the frames in supporting and travelling with the locomotive were carried by the boiler itself. In an effort to reduce the amount of metal (and therefore cost) involved in the construction of their locomotives' frames, railway companies like the London and North Western Railway (L&NWR) used the boiler structure as a 'strong-back' above their often flimsy frames to keep the locomotive straight and true; this was used in locomotives like their eight coupled freight engines during the late 19th century. This practice was discontinued in

later designs as inappropriate and unsafe.

The frames of a locomotive have to bear the pulling and pushing forces when a locomotive is working hard. The stepped shape of *Rocket*'s frames was not conducive to resisting crushing or stretching forces. *Rocket* had several accidents in its early days. Most of these shunts involved running into other vehicles. The impacts caused the frames to be compressed. To this day it is possible to see that the front end of the original *Rocket* in the Science Museum droops from the impact of the forces to which it was subjected as a hard-working locomotive.

ABOVE 2010 replica new frames being erected at the Flour Mill in the Forest of Dean. *(Flour Mill)*

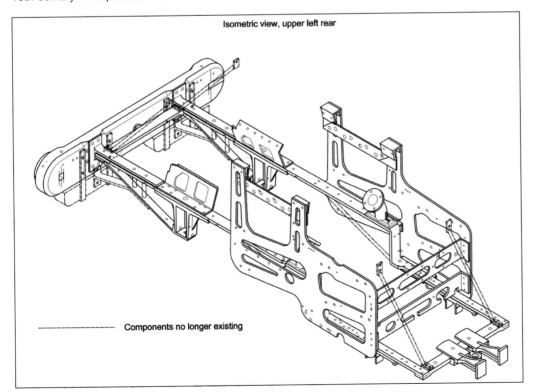

Isometric view, upper left rear

- - - - - - - - - Components no longer existing

LEFT Isometric drawing of *Rocket*'s frames showing cylinder-mounting brackets in revised position. *(John Glithero)*

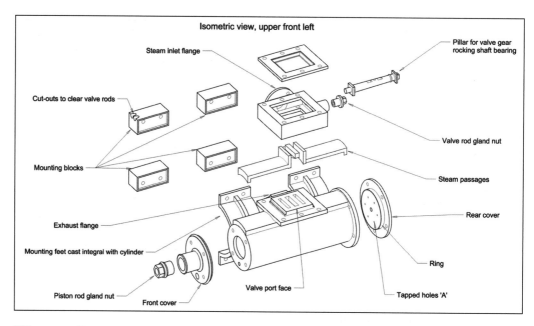

Isometric view, upper front left

Steam inlet flange

Pillar for valve gear rocking shaft bearing

Cut-outs to clear valve rods

Valve rod gland nut

Mounting blocks

Steam passages

Exhaust flange

Rear cover

Mounting feet cast integral with cylinder

Ring

Piston rod gland nut

Front cover

Valve port face

Tapped holes 'A'

The cylinders

The cylinders of a steam locomotive are at the heart of the machine and determine the success or otherwise of its design. The original position and size of *Rocket*'s cylinders closely follow the pattern set for *Lancashire Witch*, built a year earlier. There were two cast iron cylinders with 8in inside-diameter bores, positioned on iron plates attached to the frames. In the original design the cylinders were inclined at 38° to the horizontal. The pistons inside the cylinders pushed and pulled the crosshead, which was guided on square steel slide bars to reciprocate with a 16.5in stroke.

One of the big differences between *Rocket* as it was at Rainhill and *Rocket* as it exists now is that the cylinders have been brought down from being steeply inclined to the horizontal to being nearly horizontal. The effect of the steeply inclined cylinders had caused a rocking action as the locomotive worked hard. This effect was almost eliminated by slanting the cylinders nearer to horizontal. The Stephensons' understanding of the significance of this improvement can be seen as early as 1830:

RIGHT By bringing the cylinders down to almost horizontal the vertical component was all but eliminated, causing *Rocket* to ride more steadily.

BELOW Isometric view of *Rocket* left-hand cylinder and steam chest assembly. *(John Glithero)*

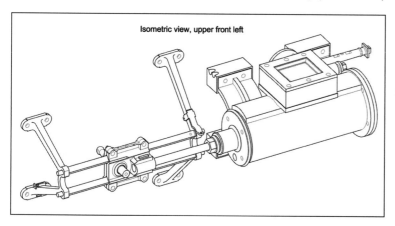

Isometric view, upper front left

Rocket's sister locomotive *Northumbrian* had near-horizontal cylinders when new.

Each steam cylinder in *Rocket* has cast into it internal steam passages leading from each end, as seen in the illustrations on the right. These passages surface on to a flat valve face approximately midway along the outside of the cylinder. This valve face was originally on the lower side of *Rocket*'s cylinders, but the diagram shows the cylinder with the valve face on the upper side, as it is in the final design. The valve face is enclosed by a box called the 'steam chest', which is fed with high-pressure steam from the boiler. Inside the steam chest, a reciprocating flat rectangular valve (driven from the locomotive axle via an intermediate rocking shaft under the boiler) spans the valve face. It is capable of covering all three ports when the valve is in the central position. As the valve moves back and forth it uncovers each end port leading to the passages within the cylinder. Steam from within the valve chamber is thus admitted to the relevant end of the cylinder at the right time. In addition the underside of the valve has a cavity that connects the central port of the valve face with each of the end passages in turn. This central port connects to the chimney to allow spent steam to exhaust. When one end of the cylinder is open to pressurised steam to push the piston, the other end is open to exhaust, and vice versa. By this means the used steam from the non-powered side of the piston can escape to the atmosphere up the chimney with the familiar *chuff* sound we associate with steam locomotives. In the illustrations on the right, the cycle of operation determined by the position of the valve and piston is shown in five consecutive positions, each representing one-fifth of a complete wheel revolution. The high-pressure steam is shown in red and the low-pressure (exhausting) steam is shown in blue.

The position and direction of travel of the valve determines whether the piston rod pushes or pulls on the wheel. The valve gear and its behaviour will be described in the section on motion. The piston is sealed into the cylinder bore by two split piston rings to maintain a steam-tight fit. The cylinder ends are closed by steam-tight bolted covers, the lower one of which has a sliding seal called a 'stuffing box' or 'gland', to allow the piston rod to pass in and out of the cylinder without leakage.

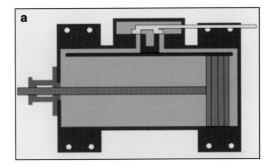

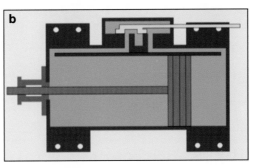

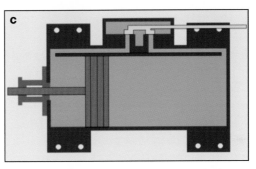

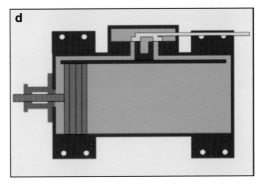

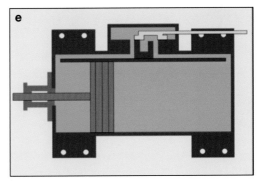

LEFT Five animated sequential views of cylinder valve and piston under steam. Steam is shown red and exhaust is shown blue. Each image represents one-fifth of a total cycle.
(Crispin Cousins)

ABOVE *Locomotion* of 1825, built by George Stephenson for the Stockton and Darlington Railway, showing the two-piece plug wheels, which were heavy but fit for purpose.

BELOW Drawing of *Rocket* wooden driving-wheel assembly, showing the various components held tightly together by the shrunk-on tyre and the iron collar to strengthen the crank boss. *(John Glithero)*

The wheels and suspension

The rules of the Rainhill Trials allowed for the competing locomotives to have more than four wheels, but the designated load to be hauled was proportionately greater with more wheels. The Stephensons' design criterion was to save weight wherever possible to reap the benefit of a lightweight load to pull during the test. They must have discussed the possibility of replicating *Locomotion*'s 1825 composite wheels but realised that the clumsy two-piece design would be too heavy for the four-wheel lightweight model they designed for the Trials.

The wheel design they settled on was wooden with an iron tyre shrunk on in the manner of a farm cartwheel, as shown below. The density of iron is 8.0gm/cc, compared with 0.8gm/cc for oak, so there was a great

potential saving of weight. But would saving weight by using wood be achieved at the expense of wheel strength when one of the spokes was being alternately pulled and pushed by the piston to drive the wheel round? To add strength, an iron strap was fitted that transmitted the piston's force on to the wheel rim and tyre where it was needed. This is shown in the illustration on page 22.

The last operation in building a heavy-duty wooden cartwheel is to shrink a red-hot iron tyre (which in this case would have been rolled to shape with a flange) on to the outer rim of the assembly. The hot tyre was quenched by dousing it with water so that it shrank on to the assembled wheel components and pulled them tightly together. This put the tyre under considerable tension around the hoop, but it was strong enough to resist the stresses exerted as the vehicle was moving.

When the 1979 *Rocket* replica was created by Mike Satow for the Science Museum it had wooden wheels as per the original, and steel tyres created using a traditional wheelwright's shrink fit. However, the replica's wooden wheels were found in service to be not as tightly clamped by the shrink-fitted outer ring as had been hoped. This meant that they never worked satisfactorily and had difficulty moving over point-work and tracks where the gauge was slightly wider than perhaps it should be. They were eventually replaced by modern heavy cast steel spoked wheels. The original wooden wheels and tyres still rest on the shelves in the National Railway Museum's store as evidence of the sad conclusion that we are not as capable of producing perfect wooden wheels as the wheelwrights of 1828.

The 56.5in diameter driving wheels were attached to the 3.25in diameter wrought iron front axle of *Rocket* by keys inserted from outside to ensure strength of drive and that they ran truly. The cranks were set at 90° so that there was always one cylinder capable of maximum effort whatever the position of the driving axle. During one of the several rebuilds in the Newcastle Works referred to previously, *Rocket*'s axle was increased to 4in diameter to provide greater strength.

Rocket's rear wheels of 32.5in diameter were fitted under the firebox. They were made of

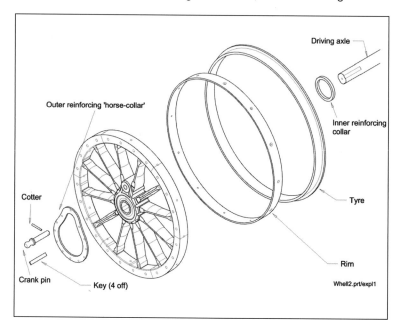

Outer reinforcing 'horse-collar'

Driving axle

Inner reinforcing collar

Cotter

Tyre

Crank pin

Rim

Key (4 off)

Whell2.prt/expl1

ABOVE *Rocket* replica driving wheel found to be unsteady at the *Rocket 150* cavalcade so replaced by steel castings, now in store at the National Railway Museum.

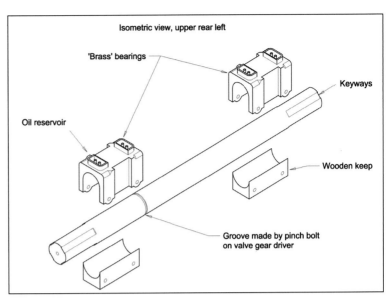

ABOVE Isometric drawing showing axle, bronze bearings and under-keeps. Note wells on top of axle boxes for lubrication. *(John Glithero)*

BELOW Front axle and spring assembly showing how the load is transferred to the frames from the wheels and axles through the springs. *(John Glithero)*

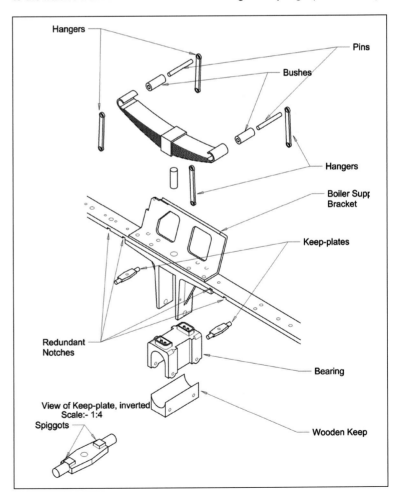

chilled cast iron and were waggon-type spoked wheels typical of the period. It is noted in Dr Michael Bailey and Dr John Glithero's work that John Rastrick recorded that the trailing wheels were *missing* the day before the Rainhill Trials, for unspecified reasons, and were substituted at the last minute by a pair of slightly larger conventional waggon wheels conveniently to hand. Perhaps this was a last-minute attempt to save weight before the weigh-in on the big day? We shall perhaps never know.

The axle shown above right was carried in bronze axle boxes that were constrained in vertical iron guides attached to the frames. These guides, known as 'horns', allowed the axle boxes individual vertical freedom, under control of the leaf springs, to cope with track undulations so that the locomotive applied reasonably even pressure over its two pairs of wheels. The sloping cylinders and pistons of the original design allowed some of the sloping drive forces to rock the locomotive from side to side as the vertical component of the driving force varied at each reversal of the piston.

After learning from the experience of this rocking movement, the redesign almost eliminated the vertical component of the driving forces. That is why we see *Rocket*'s cylinders at a lower slope in post-Rainhill state. This was cleverly done by swapping the cylinders from side to side and inverting them so that everything would fit compactly into the redesigned layout.

The motion

'Motion' is the name given to all the various moving components that are needed to transmit the drive from the pistons to the wheels, as well as to drive the valves, to enable the locomotive to become self-acting and its mechanism to operate efficiently. As well as driving the wheels round, the motion imparts axial reciprocating movement to the slide valve, which in turn distributes the live and exhaust steam at the right time to the right place. This valve motion is derived from the wheels and driving axle via a series of eccentrics and links.

From James Watt's invention of the rotative double-acting steam engine in the 1770s, the concept of the steam engine found application wherever machinery needed to be powered. However, there were at least three applications where there was a need for the machinery to be reversed and able to run backwards as well as forwards. These were ships' engines, colliery winding engines and railway locomotives. This was a problem for the early engineers. A clever method of being able to run an engine in either direction is to use a device called a 'slip eccentric' to drive the valves, to admit the steam into the right position at the right time. *Rocket* was equipped with this gear, extensively used nowadays for miniature live-steam garden railway locomotives.

In this system the steam slide valve

described previously is driven to reciprocate back and forth from an eccentric, driven from the main engine shaft, so that the steam ports open and close at the right points to drive the engine forwards. However, the eccentric is not attached rigidly to the engine shaft. It is driven by a peg on the shaft that engages with the eccentric and drives it round in the forward direction. But if the shaft rotation is somehow reversed, the eccentric will 'slip' round the engine shaft until the peg picks it up once again and starts to drive it in the opposite direction. By arranging the two places where the eccentric picks up the drive from the peg, at the correct places for forward and backward motion, the engine is able to move both backwards and forwards.

There is, however, a problem. In order to change direction, the engine needs to rotate initially – for a portion of a revolution – in the opposite direction from which it is presently set in order to get its eccentric to the alternative position. With a small model locomotive this is not a problem. A simple push of the stationary engine in the required direction flicks the eccentric to the alternative position. However, this is not a practical proposition for a full-size locomotive!

Rocket had a cylinder on each side of the engine and they were 90° out of phase. This meant that it needed two slip eccentrics also set 90° out of phase with each other but fastened together in a cluster to allow

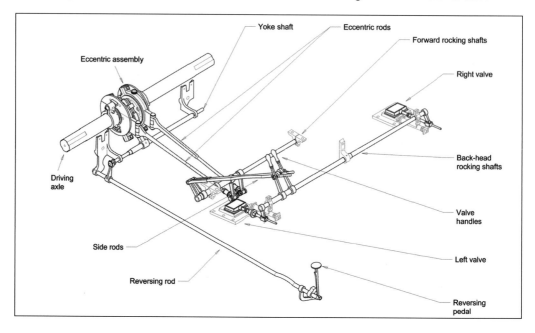

RIGHT Layout of the motion driving the valves from the front axle. *(John Glithero)*

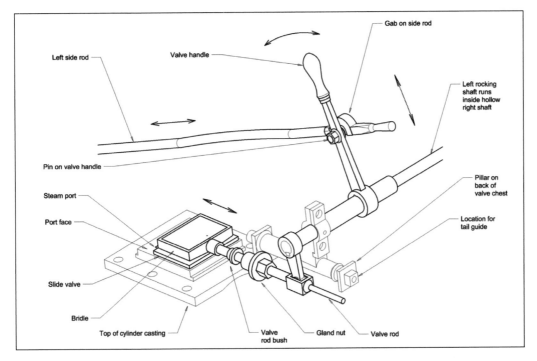

Labels on diagram:
Gab on side rod
Left side rod
Valve handle
Left rocking shaft runs inside hollow right shaft
Pin on valve handle
Pillar on back of valve chest
Steam port
Location for tail guide
Port face
Slide valve
Bridle
Top of cylinder casting
Valve rod bush
Gland nut
Valve rod

LEFT Valve gear gab action, showing how valve linkage can be disconnected for manual control of the valves at starting or reversing. *(John Glithero)*

them to 'slip' around the driving axle together during reversing. The direction of eccentric slip required to reverse the engine was in fact the same direction as the locomotive was required to set off. This is an impossible ask – because that forward movement was not possible until the locomotive was fully reversed. The early steam engineers overcame this by providing those driving the engine with the ability to make the first counter-revolution by separate hand controls called 'gabs', which could be uncoupled to put the steam directly into the cylinder where it was wanted, independently of the valve gear. Once the machine had moved in the new direction sufficiently for the eccentric to pick up the drive again in reverse, the engine could sustain its motion automatically.

The really magic breakthrough that Robert Stephenson incorporated into *Rocket* was not to expect the driver to start the locomotive by hand, but to allow the locomotive to drop into reverse whilst it was still rolling forward in preparation for stopping once steam had been shut off. This was a masterstroke, yet was simply achieved. Instead of the eccentrics alternating between two fixed points on the driving axle for forward and reverse, he provided two clutching points in the rotation of the axle where the valve gear could be pre-selected for whatever the next operation was intended to be.

LEFT Eccentric cluster on front axle of *Rocket*. Both driving plates can be seen with the rectangular holes for the 'dog' to drop into when a particular side is selected. The dog can be seen partially engaged into the far side slot.

Of course, sceptical readers will have already realised that there is a fatal flaw in all this cleverness. Let us imagine the train rolls towards the terminus buffer-stops ready to set out later in reverse, and that the driver has sensibly pre-selected reverse before the train comes to rest. Someone in authority then perhaps asks the driver to pull forward a small amount ... which is simply not possible without resorting to hand control to override the slide valves' natural positioning.

Nevertheless, the reversing gear that Robert Stephenson settled on was ideally suited to the task of reversing. He was rightly proud of how it was designed and worked out. He said at the time, in one of his letters available in the National Railway Museum's archives, 'This hand gear on one of the locomotives at Liverpool answers as well as anything possibly can do,

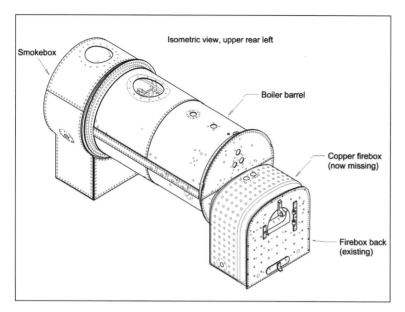

Isometric view, upper rear left

Smokebox

Boiler barrel

Copper firebox
(now missing)

Firebox back
(existing)

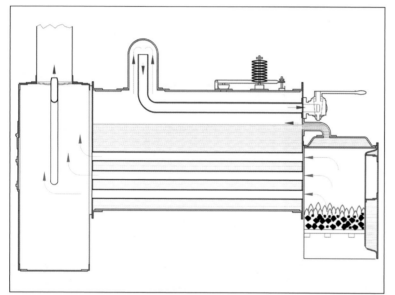

and the men like it very much.' He also wrote to Henry Booth, 'I expect the mode for changing the [valve] gear will please you, it is now as simple as I can make it and I believe effectual.'

It was not until much later, in 1845, that one of George Stephenson's employees called William Howe, working for the Clay Cross Company in Derbyshire, managed to create what we would now recognise as 'Stephenson's link motion' for proper reversing. Not only was the engine easy to reverse on the move, but also the working of the engine proved to be much more economical in the intermediate positions of the eccentrics between forward and reverse. By being able to reduce the valve travel the quantity of steam being used on each stroke was lessened to achieve the same effect. The prototype steam engine on which William Howe carried out that important development work is on display at Kelham Island Museum in Sheffield.

The boiler and related systems

Rocket's boiler and firebox were at the heart of its groundbreaking success. The boiler barrel was made up of a drum shape, rolled into a cylinder from flat sheets of 0.25in thick wrought iron and riveted along the seams. It had rolled angle iron rings at each end to enable the flat end plates to be riveted on to close the drum. In the lower half of the drum (below the water level), 25 copper tubes, each of 3in diameter were grouped in a honeycomb pattern. These conducted the flames and heat from the fire through the space where the water was.

The external firebox, encased by water, was made of 0.25in thick copper sheet. The firebox was attached to the rear of the boiler in the tube area. The chimney base was attached to the front

LEFT Inside the tube space of *Rocket*'s boiler.

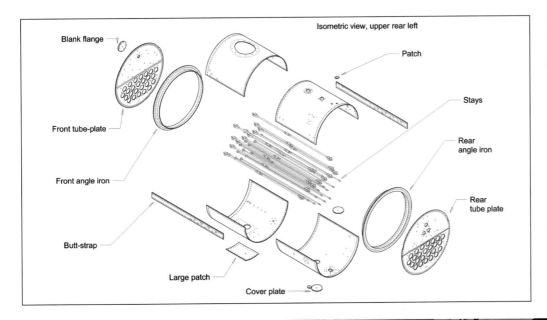

Isometric view, upper rear left

Blank flange

Front tube-plate

Front angle iron

Butt-strap

Large patch

Cover plate

Patch

Stays

Rear angle iron

Rear tube plate

LEFT Exploded view of *Rocket*'s boiler. *(John Glithero)*

BELOW Inside the tube space of the replica *Rocket*'s boiler, showing how the internal fittings are positioned.

of the tube area. The chimney base turned the gases and flames from the fire upwards, where the blast pipes from the engine's exhaust steam were situated. Inevitably the blast encouraged ash to be drawn through the boiler. It was the lack of capacity of this small smokebox at the front that caused problems later on. After a year of operation the front end of *Rocket*'s boiler was

BELOW The boiler remains of *Rocket* viewed looking towards the firebox, to the three holes where the regulator and the water circulating pipes attach on the rear tube-plate.

BELOW RIGHT Looking into the sectioned firebox of the *Rocket* replica.

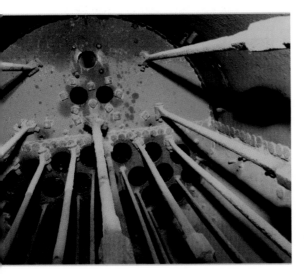

rebuilt, into the shape shown in the final form seen in the illustration at the top of page 56. This meant that there was ample capacity for the cinders and char from the fire that were thrown forward by the blast, without impeding the locomotive's performance by requiring pauses to clear the debris away partway through the journey between Liverpool and Manchester and back.

The firebox with its water jacket had to allow the water inside it to circulate freely around the rest of the boiler, so four large copper circulating pipes joined the copper firebox to the circular barrel. These features are indicated in John Rastrick's sketches (below). Rastrick was a well-established and respected steam engineer and had been selected to be one of the Rainhill Trials judges, along with Nicholas Wood and John Kennedy. It is fortunate that he sketched and wrote prolific notes of what he observed as a judge. For example, his sketches also show the multitude of stays built into the copper firebox to prevent the flat sides from being pushed apart by the internal pressure.

For the Rainhill Trials, the weight limit imposed in the competition specification challenged the Stephensons to create an innovative boiler for *Rocket*. Many engineers had previously designed and built static boilers that incorporated the heavy weight of the metalwork surrounding the water being boiled. Indeed, two of the competitors' locomotives at

Rainhill were built on those principles. Heavy static boilers were thermally more stable because the mass of metal surrounding the water gave up its heat to compensate for fluctuations in hand firing. With Henry Booth's multiple tubes, *Rocket*'s boiler broke new ground. The nest of copper tubes inside the water space meant that the total amount of water in the boiler was greatly reduced because the tubes were taking up space normally occupied by water. That in itself saved weight. Moreover the boiler in *Rocket* was far more responsive to change in steam demand, as the fire boiled the reduced quantity of water to steam more quickly.

Previous to *Rocket* boiler designs were often built with substantial, dished, wrought iron plate ends, like those we see on *Puffing Billy* or *Locomotion*. The single fire-tube inside was fastened into the ends. The fire-tube braced the end plates as the pressure inside the boiler pushed outwards on to them. With no strong central tube, the Stephensons had to work out a new way of sealing the 25 copper tubes into the end plates, as well as providing additional support for the flat end plates themselves. This challenge caused Robert some head scratching. He chose to seal the tubes into the boiler end plates or tube-plate using a method called 'clinking', which we would now call expanding. A tapered iron or steel ring or

RIGHT John Rastrick's notebook with sketches and notes of *Rocket*'s firebox and boiler at the time of the Rainhill Trials.

ferrule was fitted tightly into the copper tube end where it projected beyond the tube-plate. The ferrule was pressed into the end of the tube to swell out the copper so that it was tightly locked into the tube-plates by the distortion of the tube wall. The ferrule was left in place inside the copper tube at the points where it was necessary for the tube to be sealed into the tube-plates (above and above right).

The word 'clinking' that Robert Stephenson used in his correspondence is fascinating. He even used the past tense – 'The tubes are all clunk into the boiler…'. It is not a word that we use nowadays, but the technique is well described in the contemporary accounts of boiler making. Perhaps it was the ringing sound as the iron ferrules were hammered home within the tubes that inspired the term. Certainly as the ring expanded the copper tubes in the confined space the copper would work-harden and increasingly resist the impact on the ferrule as it was hammered into place. But that was not the end of Robert Stephenson's assembly problems with the new boiler.

The Rainhill Trials rules initially stated that the finished boilers must be subjected to three times the working pressure on hydraulic test. That is 150psi. Perhaps Robert had thought that the 25 tubes with their clinked ends would give more than sufficient bracing to the end plates. However, on hydraulic test the tube-plate started to bow out and distort under pressure. Robert Stephenson wrote to Henry Booth to report progress on 26 August 1829, as the deadline loomed for the Trial:

'On Wednesday I had the boiler filled with water and put up the pressure of 70lb per square inch when I found that the yielding of the boiler end injured the clinking of the tubes. I therefore thought it prudent to stop the experiment until we got some stays put into the boiler longitudinally. The boiler end at 70 lb per sq. inch came out a full 3/16 of an inch. This you may easily conceive put a serious strain on the clinking of the tubes.'

The original letter, retained in the archives of the L&NWR, was destroyed by fire when on loan to the Brussels Exhibition of 1910. Fortunately a photographed copy is held in the National Archives (008/88/1.4).

The stays were a series of iron rods with clevises at the ends fastened firmly to both boiler end plates. The rods were fitted with left- and right-hand threaded hexagonal turnbuckles at about the mid-point to adjust and tension their length without disturbing the end fittings. In the photographs of the sectioned replica this may look like an easy task, but in fact it would

ABOVE Adjustable stays on the sectioned replica, showing turnbuckles with left- and right-hand threads to adjust the tension on the stays. (Author)

RIGHT Boiler water level gauge on 1935 Rocket replica. A drain valve is fitted at the bottom of the gauge to verify a true level. (Author)

BELOW Twin try cocks on 1935 replica at boiler water level. Results from these should be treated with caution, as indicated in the text. (Author)

have to have been carried out inside the closed boiler by lying on top of the barrel and reaching down into the dark with both arms to tighten the turnbuckles. These are also seen in John Rastrick's sketches.

To ensure safety, anyone who uses a boiler to heat water should know about the importance of maintaining an adequate water level. This applies whether the boiler in question is a kettle for making tea or a steam locomotive boiler powering Rocket. A perfectly safe and correctly maintained steam boiler can become a potential bomb if the correct safe water level is not maintained properly at all times. Rocket's boiler was equipped with various auxiliary systems that enabled this condition to be met satisfactorily. These are described below.

The water gauges

As with a modern electric kettle, the Stephensons' design had a transparent glass tube water gauge. It was positioned on the side of the circular barrel straddling the correct water level for the boiler. In addition the boiler was fitted with two 'try cocks' – separate brass valves fastened into the boiler shell close to where the water level should be. To check the boiler water level, the fireman used the try cocks by opening each valve in turn to watch what came out of them. It seems a simple, foolproof safety device, but in reality it gave unreliable results. Steam came out of both cocks even if the water level was correct! The original try cocks are missing on Rocket, but they are shown on the replica below left.

When the higher of the valves is opened, steam should emit confirming that the level is below that of the highest valve. So far so good. But when the lower valve is opened, which should be below the water level, steam still comes out! This is because the boiling water is under pressure (which elevates its boiling point) and is likely to be boiling at about 130°C. What comes out of the lower valve is steam, not water, because the water (well above boiling point) immediately flashes off to steam. This can cause confusion for inexperienced locomotive crew who may be expecting to see water when the lower try cocks are opened.

LEFT **Crosshead-driven water feed pump on 1935 replica.** *(Author)*

FAR LEFT **Drive to crosshead-driven feed pump on 1935 replica.** *(Author)*

LEFT **Crosshead-driven water feed pump pipework feeding water to the boiler from the tender, via the shut-off valve.** *(Author)*

The feed pump

*R*ocket was fitted with one (possibly two) mechanical water feed pump(s) driven from the crosshead of the locomotive. The ram of the pump was 1.25in in diameter. This forced water into the boiler through a non-return valve at each piston stroke. The purpose of the feed pump was to replenish the small amount of water from the boiler that was used as steam with every working stroke of the pistons, as they drove the wheels round. The prominent pipework above the right-hand

cylinder attached to the pump was to enable the pumped water to be cycled round the pumping circuit on demand, so that it was not forced into the boiler when the boiler was already adequately full. The pump drew fresh water from an adapted (empty!) sherry barrel, mounted on the tender, via a flexible pipe and water shut-off valve between engine and tender. There might have been a hand-operated force pump mounted under the tender barrel to put water from the barrel into the boiler by hand when necessary, but no evidence exists to support this theory. Later steam locomotives

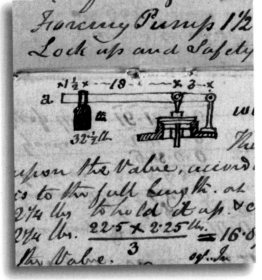

ABOVE 1979 *Rocket* replica blowing-off surplus steam through the safety valve at the National Railway Museum, York.

RIGHT Page from John Rastrick's Rainhill Trials notebook, showing safety valve on *Rocket*.

BELOW Deadweight and Hackworth spring safety valves on sectioned 1935 replica. *(Author)*

fitted with crosshead pumps like *Rocket* were known to have the capacity to pump water into their boilers when stationary by means of oiling under the driving wheels and opening the throttle to spin them, taking care to ensure the locomotive could not move.

The safety valves

Steam boilers must have safety valves capable of discharging any over-pressure steam being produced by the boiler at the highest possible steam production rate. For enginemen struggling to get their locomotives up hills whilst steam wasted itself through the safety valves, the temptation to interfere with the safety device to get a bit more pressure was enormous. Several engine crews lost their lives in the early days of steam through explosions caused by holding the safety valve down to get more power from the engine. Of the two safety valves fitted to *Rocket*, one was a spring type padlocked to prevent interference by over-zealous drivers. The other was a weighted arm, shown on the left. This safety device was ideally suitable for a stationary boiler, but unstable if a locomotive was running along an uneven track and certainly a waster of precious steam as it escaped at each jolt. This is another example of technology from equipment suitable for

LEFT Dome on boiler of 1935 sectioned replica. *(Author)*

static machinery being found inadequate when transferred to moving vehicles.

Timothy Hackworth invented a type of spring safety valve that would not be subject to track jolting and consequent steam wastage. It consisted of a series of plate springs stacked back to back as shown at the foot of page 62.

The dome

The dome of a steam locomotive is usually a hemispherical-shaped cover situated on top of the boiler. Its purpose is to allow the steam to be drawn from a high point on the boiler that is *only* steam, and not water droplets.

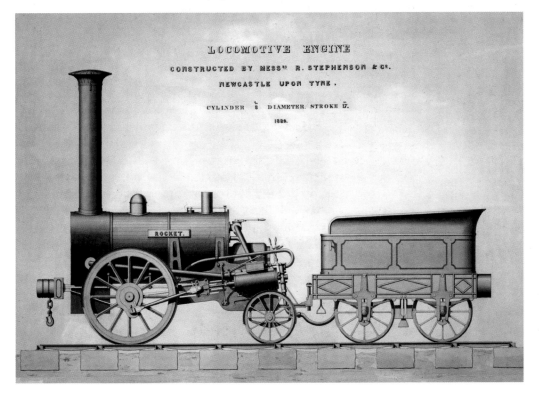

LEFT 1836 coloured general arrangement drawing of *Rocket*, known as the 'Crewe drawing', formerly in possession of the London & North Western Railway and now with the Science Museum.

The explanation of this statement is that when steam is drawn from a boiler there is a tendency for water to jump up out of the boiling water surface and follow the steam into the steam pipe to the cylinders. This effect is known as 'priming', and can cause problems for the performance and longevity of a locomotive. *Rocket* had no dome when first constructed so technically this is another area where the National Railway Museum's sectioned replica is wrong. The feature that looks like a dome in the Rainhill Trials versions of *Rocket* is in fact the second safety valve, but this was not really understood until Dr Michael Bailey and Dr John Glithero did a thorough job in establishing that priming (or water carry-over from boiler to the cylinders) became an early technical problem when *Rocket* was in steam. The problem was solved by creating a proper dome on top of the manhole situated at the front of the top of the boiler. The steam pipe collected steam from an inverted pipe at the top of the inside of the dome, as in the photograph of the replica. That excluded most of the water swilling about below the dome as the locomotive moved along the track. The safety valve that had been on the manhole was re-sited at the rear of the boiler following rebuilding; so what looks like a dome on the image of *Rocket* on the front cover of this book is in fact a lock-up safety valve.

The regulator

'**R**egulator' is the railway term for the valve that allows steam under pressure into the cylinders from a point high up in the boiler. Sometimes it is called the throttle. It determines how much steam is admitted to the cylinders and therefore how fast the engine can travel. It is similar to a large gas tap. It has a hollow taper plug which fits snugly in a tapered hole in the regulator body as shown.

The grate

The grate of *Rocket* was like an old-fashioned domestic grate, with cast-iron fire-bars resting on ledges in the open bottom of the copper firebox. The cinders and ash produced as the coke burned fell through the bars on to the track. However, a big advantage over previous

LEFT The 1935 replica *Rocket* newly constructed at Robert Stephenson & Co, Darlington.

BELOW Rainhill Trials poster of the two participants, *Rocket*, and *Sans Pareil*, showing the common tender design.

locomotives designs was that the limited space under the fire-bars did not have to be continually cleared of ash and clinker. This would have been the case with *Locomotion* or *Lancashire Witch*, where the fire and grate were confined to a tube and all the debris from previously burned fuel fell through the bars into the space below.

Strangers to the subject of steam locomotives have difficulty understanding how it is that simple cast-iron fire-bars can support a high temperature solid-fuel fire without collapsing as the fire-bars reach their melting temperature. There are two golden rules. Firstly, the air space below the bars must be kept clear of ash at all times so that the incoming cold air for combustion continually cools the bars supporting the fire. Secondly, the scrap iron content of the cast iron from which the bars are made must be high, in order to keep the melting point as high as possible.

The tender

Steam locomotives needed a tender to provide a travelling receptacle for the fuel for the fire and water for the boiler. *Rocket*'s tender was not built by the Stephensons. Bailey and Glitherblished that it was built locally in Liverpool by a firm of coach-makers called Thomas and Nathaniel Wordsell, specifically for the Rainhill Trials. Used sherry barrels must have been easily available in a port such as Liverpool at the time of the Rainhill Trials, so a simple four-wheel vehicle was designed which carried fuel and water and a platform away from the footplate for the crew.

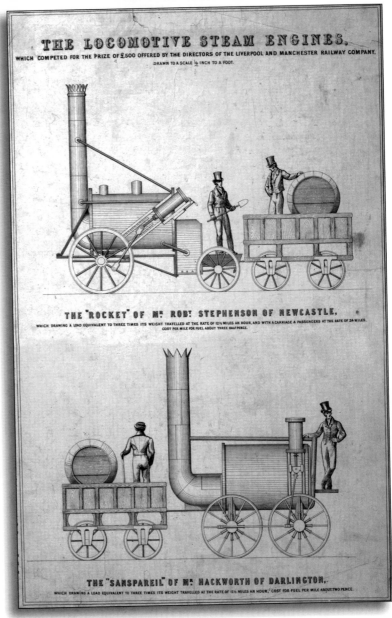

THE LOCOMOTIVE STEAM ENGINES,

WHICH COMPETED FOR THE PRIZE OF £500 OFFERED BY THE DIRECTORS OF THE LIVERPOOL AND MANCHESTER RAILWAY COMPANY.

DRAWN TO A SCALE ½ INCH TO A FOOT.

THE "ROCKET" OF M.ᵣ ROB.ᵗ STEPHENSON OF NEWCASTLE,

WHICH DRAWING A LOAD EQUIVALENT TO THREE TIMES ITS WEIGHT TRAVELLED AT THE RATE OF 12½ MILES AN HOUR, AND WITH A CARRIAGE & PASSENGERS AT THE RATE OF 24 MILES. COST PER MILE FOR FUEL ABOUT THREE HALFPENCE.

THE "SANSPAREIL" OF M.ᵣ HACKWORTH OF DARLINGTON,

WHICH DRAWING A LOAD EQUIVALENT TO THREE TIMES ITS WEIGHT TRAVELLED AT THE RATE OF 12½ MILES AN HOUR, COST FOR FUEL PER MILE ABOUT TWO PENCE.

ABOVE Replica *Rocket* and coaches showing coupling height disparity between locomotive and tender.

It is significant that the same design of tender used for *Rocket* was used by Timothy Hackworth's *Sans Pareil*, so there is almost a suggestion that the tenders were perhaps free issue for the Rainhill Trials competition and supplied by the railway company. Of course, *Novelty* carried its own coal and water within the locomotive so it did not need a tender. In spite of the obvious difference in the locomotives, it is clear that Timothy Hackworth

and George Stephenson shared a common tender design. The artist clearly thought that driving and firing steam locomotives was a gentlemanly occupation that demanded sartorial elegance. Nothing could be further from the truth!

There was a peculiarity associated with the tender supplied to travel with *Rocket* that only became an issue when *Rocket* was connected to carriages with conventional-height buffers and couplings. Looking more carefully at the image of *Rocket* at the top of this page, the coupling to the locomotive is arranged at below footplate level height, as the locomotive's draw-gear to pull its train is attached to the rear of the locomotive's frame. However, the actual coupling to pull the train at the rear of the tender is arranged some 18in higher than the coupling at the front. This disparity in coupling heights meant that under heavy pulling load conditions the line of the couplings would try and straighten itself out, causing the tender to rotate and derail.

LEFT *Locomotion*'s driver position in 1825, on top of the boiler in order to be able to control the valve gear. A centenary poster for the LNER.

The footplate and fittings

It might seem strange to the reader to classify a simple flat metal plate at the rear of the boiler of *Rocket* as part of its anatomy, but in fact this simple plate, about 4ft square, represented a significant development that was first seen in locomotive design about the time of *Rocket*. It was to remain with UK steam locomotives for the next 130 years.

In contrast, *Locomotion* – designed by George Stephenson some five years earlier – was driven from a platform alongside the cylinders, as shown at the bottom of page 66. Although the driver of *Locomotion* had good forward vision, he was clearly in an unsafe position to do his work, sitting atop a hot, slippery boiler, surrounded by flying machinery, and working with his back to his fireman, with whom he needed to communicate well in order

LEFT *Sans Pareil*'s replica driver position makes communication with the fireman somewhat tricky. *(Martyn Stevens BBC)*

ABOVE Locomotive-to-tender coupling position on *Rocket*.
(John Glithero)

step forward, from both safety and ergonomic viewpoints. The locomotive's operating controls, as well as access to the firebox, fuel and water supply, were at last properly positioned. The footplate of *Rocket* rested on the frame extension at the rear and covered a coupling socket which extended rearwards and to which *Rocket*'s tender was attached.

However, the most important fitting to be found on *Rocket*'s footplate was an insignificant-looking pedal that projected up through the plate on the left-hand side close to the rear of the firebox. This pedal was to be the undoing of many overconfident *Rocket* drivers on both the original *Rocket*, no doubt, and the working replicas.

The pedal was positioned there to slide the pair of eccentrics on the front axle from the forward to the reverse position and vice versa. It was spring loaded so that it defaulted to the forward condition. There was a position between the two driving plates where the eccentric cluster was in neither forward nor reverse gear. When initiating the selection of reverse gear, the footplate person had to apply their heel to the pedal with all their weight and bounce on it to latch it against the spring into the reverse condition, then wait for a part of a revolution of the wheels until it dropped into the chosen place. Trying to achieve this with slightly built and inexperienced footplate crews on the replica *Rocket* provided hilarious entertainment to onlookers, and I have no reason to doubt this was the same challenge for drivers of the original *Rocket*. This issue is discussed more fully in chapter three in the section on how to drive *Rocket*.

to do his job properly! The fireman was in the safest position but he had to anticipate the driver's intentions.

Timothy Hackworth's *Sans Pareil* showed that by 1829 there had been little ergonomic progress in footplate design. The firing and driving stations were positioned at opposite ends of the locomotive, and the prerequisite of the crew to be in good communication with each other was not fulfilled in such an arrangement. So the grouping together of driver and fireman on a forward-facing footplate that allowed the crew both to see forwards and communicate with each other was a positive

BELOW Footplate reversing pedal on 1935 *Rocket* replica.

The paintwork

The paint scheme chosen is an important part of the legacy of the anatomy of *Rocket*. The colours selected were cleverly chosen by the Stephensons to send a clear message to their audiences in the same way that a bee or a wasp sends an instant visual message to observers: *Do not mess with me! I look good and I mean business!*

Horse-drawn stagecoaches were usually painted in bright attractive colours and were given names that suggested speed and

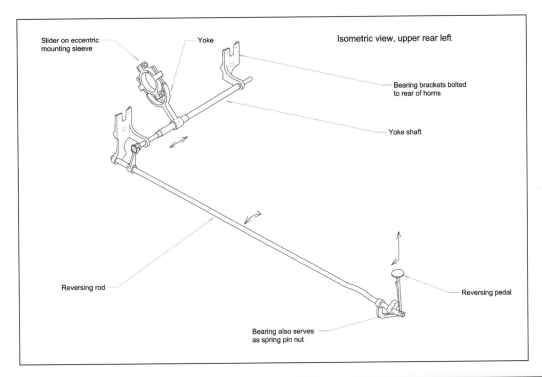

LEFT Layout of the motion driving the valves from the slip eccentrics on the front axle. *(John Glithero)*

Slider on eccentric mounting sleeve

Yoke

Isometric view, upper rear left

Bearing brackets bolted to rear of horns

Yoke shaft

Reversing rod

Reversing pedal

Bearing also serves as spring pin nut

BELOW *Rocket* replica 2010 publicity image.

efficiency. Light colours like white and sunflower yellow were not easy shades to deal with for a steam locomotive, where oil and dirt are part of the daily routine. *Rocket*'s vivid yellow body with white chimney and black working parts was designed to win hearts and minds at the Rainhill Trials, not with practicalities of maintenance in mind. *Rocket*'s startling livery did engage the crowds, but only after their brief love affair with *Novelty* ended.

Robert Stephenson and his team painted *Rocket* as he was assembling the locomotive. In one of his letters to Henry Booth in August 1829, copies of which are in the National Railway Museum archive, he commented, 'The wheels of the engine are painted in the same manner as coach wheels and look extremely well. The same character of painting I intend keeping up. The engine will look light which is one object we ought to aim at.'

It is hard for us now to look at the remains of *Rocket* and to conjure up the excitement that was generated in the public's minds when the machine was brand sparkling new. The fortunate thing is that by having a working replica able to travel the world, as well as the real original machine in its developed condition in a museum, we can let the excitement read across from the replica to the original *Rocket* and dream of how it must have been in 1829.

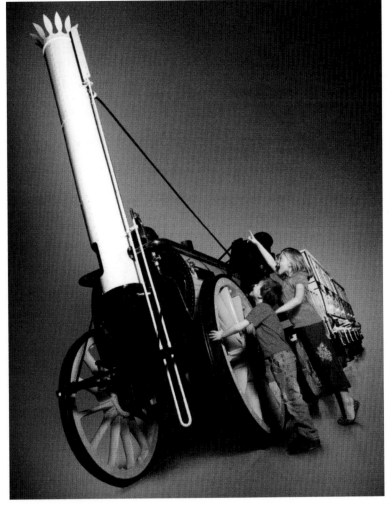

CHAPTER THREE

Driving, firing and riding on *Rocket*

'Excitement at seeing Rocket at Rainhill.' This Alan Fearnley
painting shows us the skew bridge at Rainhill as well as
providing good detail of the Birkinshaw-type wrought iron
track. It was painted for the *Rocket 150* celebrations in 1980.
(Rainhill Railway and Heritage Society)

Before we can get a proper idea of what it was like to experience travelling with *Rocket* first hand, it is important to understand the context of travel between Liverpool and Manchester in the early part of the 19th century.

Horse-drawn stagecoaches, typically carrying up to ten persons, were the established way for people to travel between the cities. Heavy goods were carried on horse-drawn barges along the Bridgewater Canal at walking pace. The invention of the macadamised road, when they replaced the poorly surfaced roads of the time in the 1820s, helped the stagecoaches increase their speed, but the coming of the railway threatened the future of both stagecoaches and canals as viable modes of transport.

There was particularly stiff competition from the stagecoach business carrying passengers between Liverpool and Manchester. Eight to ten stagecoaches were running continually between those places (see Joseph Ballard, *England in 1815 as seen by a young Boston Merchant*, published by Houghton Mifflin in 1913). With an average speed, including stops, of 8mph it is easy to see why the directors of the L&MR set the target overall speed for the Rainhill Trials locomotives with their payloads at 10mph.

There was a two-month gap between the time of the Rainhill Trials and the redeployment of *Rocket* to the construction

BELOW The canal journey between Liverpool and Manchester often took cotton longer to arrive than the time it took for it to cross the Atlantic Ocean from America.

of the remainder of the L&MR whilst the track was being completed from Rainhill onwards to Manchester. Visitors came from far and wide to experience seeing and riding behind *Rocket*. An advertisement was taken out in the *Liverpool Mercury* of 30 October 1829, a fortnight after the Trials, stating:

'ANOTHER EXHIBITION OF
MR STEPHENSON'S STEAM CARRIAGE
THE ROCKET
Will take place at RAINHILL, THIS
AFTERNOON, at ONE O'CLOCK when, as
we presume, some further EXPERIMENTS
will be made on the RAILROAD
The fact of the Exhibition may be implicitly
relied on.'

Driving, firing and riding with the original *Rocket*

Three accounts survive of the extraordinary experience of being on the footplate or riding behind the Stephensons' *Rocket* in 1829. The accounts come from sources with very different standpoints. One is from a young actress working in a theatre in Liverpool, who by the end of the trip on *Rocket* declared that she had fallen hopelessly in love with George Stephenson and his apparent charms. The second is from a vociferous opponent of steam-powered railway travel, one Thomas Creevey MP. The third is from a remarkable near-centenarian who recounted his experience of being recruited as driver and fireman on *Rocket* for the L&MR's opening day. The chapter closes with an amusing recent account from one of the regular drivers of the 1979 National Railway Museum *Rocket* replica.

Fanny Kemble's account

In her journal *Record of a Girlhood*, Fanny Kemble could not help using horse-drawn stagecoach analogies to describe her ride on *Rocket*. This account is given in full, as it provides such a powerful and authentic description of her experience:

'We were introduced to the little engine which was to drag us along the rails. She (for they

make these curious little fire horses all mares) consisted of a boiler, a stove, a platform, a bench, and behind the bench a barrel containing enough water to prevent her being thirsty for fifteen miles, the whole machine not bigger than a common fire engine. She goes upon two wheels, which are her feet, and are moved by bright steel legs called pistons; these are propelled by steam, and in proportion as more steam is applied to the upper extremities (the hip-joints, I suppose) of these pistons, the faster they move the wheels; and when it is desirable to diminish the speed, the steam, which unless suffered to escape would burst the boiler, evaporates through a safety valve into the air. The reins, bit, and bridle of this wonderful beast, is a small steel handle, which applies or withdraws the steam from its legs or pistons, so that a child might manage it. The coals, which are its oats, were under the bench, and there was a small glass tube affixed to the boiler, with water in it, which indicates by its fullness or emptiness when the creature wants water, which is immediately conveyed to it from its reservoirs.

'This snorting little animal, which I felt rather inclined to pat, was then harnessed to our carriage, and Mr. Stephenson having taken me on the bench of the engine with him, we started at about ten miles an hour ... [George Stephenson's] way of explaining himself is peculiar, but very striking, and I understood, without difficulty, all that he said to me ... The engine having received its supply of water, the carriage was placed behind it, for it cannot turn, and was set off at its utmost speed, thirty-five miles an hour, swifter than a bird flies (for they tried the experiment with a snipe). You cannot conceive what that sensation of cutting the air was; the motion is as smooth as possible, too. I could either have read or written; and as it was, I stood up, and with my bonnet off "drank the air before me." The wind, which was strong, or perhaps the force of our own thrusting against it, absolutely weighed my eyelids down. When I closed my eyes this sensation of flying was quite delightful, and strange beyond description; yet strange as it was, I had a perfect sense of security, and not the slightest fear.

'Now for a word or two about the master of all these marvels, with whom I am most horribly in love. He is a man from fifty to fifty-five years

of age; his face is fine, though careworn, and bears an expression of deep thoughtfulness; his mode of explaining his ideas is peculiar and very original, striking, and forcible; and although his accents indicates strongly his north country birth, his language has not the slightest touch of vulgarity or coarseness. He has certainly turned my head. Four years have sufficed to bring this great undertaking to an end. The railroad will be opened upon the fifteenth of next month. The Duke of Wellington is coming down to be present on the occasion, and, I suppose, what with the thousands of spectators and the novelty of the spectacle, there will never have been a scene of more striking interest.'

(Fanny Kemble, *Record of a Girlhood*, Henry Holt & Co, New York, 1883, pp158–165.)

This very early example of the 'romance of the railway' was perhaps a precursor to Britain's unique relationship with locomotives and railways. The legacy of this relationship is demonstrated with iconic films such as *Brief Encounter* and *The Railway Children*.

Thomas Creevey's account

Thomas Creevey MP, who was a vocal opponent of railways, was also taken for a ride at 23mph by George Stephenson, and wrote:

ABOVE Stagecoach travel in the early 19th century gave the L&MR a target journey time to compete with.

'I had the satisfaction, for I cannot call it a pleasure, of taking a trip of five miles in it, which we did in just quarter of an hour – that is twenty miles an hour. As accuracy on this subject was my great object, I held my watch in my hand at starting and all the time, and as it has a second hand I knew I could not be deceived. But observe, during these five miles. The machine was occasionally made to put itself out or go [indecipherable]; and then we went at the rate of twenty three miles an hour, and just with the same ease or motion or absence of friction as the other reduced pace. But the quickest motion to me is frightful; it really is flying, and it is impossible to divest yourself of the notion of instant death to all upon the least accident happening. It gave me a headache which has not left me yet.'
(Diary of Thomas Creevey MP, 1829, quoted in *The Creevey Papers*, London, 1903.)

What would Thomas Creevey have made of today's Japanese Bullet Train travelling in service at 200mph?

To put the quote above in context we should recognise the prejudices against locomotives that confronted George Stephenson and fellow builders. A stranger in the North-East, on seeing his first steam locomotive, is quoted in Samuel

Smiles' *Story of the life of George Stephenson* (published by John Murray, London, in 1862) as follows:

'The stranger had never heard of the new engine, and was almost frightened out of his senses at its approach. An uncouth monster it must have looked ... working its piston up and down like a huge arm, snorting out large blasts of steam from either nostril, and throwing out smoke and fire as it panted along. He claimed he had just encountered a terrible deevil on the high street road!'

There was a vociferous anti-railway lobby and political opposition to the Stephensons and their locomotives. There was a fear that people would expire at such high speeds, and that the sight and noise of locomotives in the fields would cause cows to stop producing milk. Although these beliefs seem laughable to us nowadays, it is important to remember that early astronauts were quarantined on their return to Earth in the 1960s in case they had somehow mutated through high-speed travel.

There are no surviving reliable, first-hand accounts of what it was like to operate *Rocket* in those heady days of late 1829 at Liverpool. The firemen and drivers would not have been expected to record what was essentially craft knowledge. Some of us with an engineering perspective have been fortunate enough to get to know the replica *Rocket* locomotive with its sparkling performance and its wilful behaviour, and contemporary accounts of operating the *Rocket* replica will be included in chapter six. It is possible by drawing on these experiences to imagine ourselves in the position of being witness to the experiences that Fanny Kemble enjoyed and Thomas Creevey hated!

A fireman's account

Let us imagine that the writer has been requested to accompany George Stephenson as his fireman on one of those heady public relations days when *Rocket* hauled a couple of carriages of excited and curious sightseers over the partially completed section of the L&MR from Liverpool to Rainhill. The fireman's account would have read something like this:

'In order to be ready for a midday departure as advertised from the terminus at Crown Street station at Edge Hill Liverpool, I had to arrive on site at seven o'clock to prepare *Rocket* for its day's work of giving many folks their chance of a first ride on a steam locomotive hauled train.

'Our *Rocket* locomotive was still warm from its previous day's exertions but I had to check that the water level in the boiler left by last evening's departing crew was at the correct level. Looking at the glass, it had to show the water level visible between the bottom and one-third full in the glass. *Rocket*'s wheels were chocked front and back with wooden wedges to ensure that no movement was possible whilst the locomotive was being prepared to raise steam.

'The remains of yesterday's fire had been allowed to die out slowly last evening. I raked the remaining ashes and clinker through the fire-bars by opening the fire-hole door and pulling and pushing with the rake to ensure all the fire-bars were clean and free to move about. The water barrel on the tender was filled from a nearby hose and tap. The two small removable doors at the front of the locomotive in the bottom of the chimney were opened up and ash and char that had gathered in that space was raked out on to the floor so that the limited smokebox space was as clear as possible.

'At the same time, the "clinked" tube ends (which could all be seen through the inspection panel as well as the firebox door), were checked with a flare light to see that there were no signs of water leaks from the tube ends or, indeed, anywhere in the firebox. The inspection doors were bolted back up and, returning to *Rocket*'s footplate, a lighted paraffin-soaked rag was dropped on to the fire-bars. Once this was properly alight some scrap pieces of dry timber kindling were thrown on to the burning rag, and in no time at all there was a vigorous fire burning in the firebox, with smoke from the fire pouring from the multi-petal-shaped chimney top. The burning timber was followed a few minutes later by a few well-chosen lumps of coal on top of the burning wood. This burned with black smoke for a while until the fire was well established and able to accept about four shovels full of coke, spread evenly about the grate, from the floor of the tender in front of the water barrel.

'At this stage local boiling sounds, like those from a domestic kettle, could be heard from the firebox even though the boiler was far from hot. I could feel the copper circulating pipes passing the warm water into the boiler barrel.

'Now there was time to clean the locomotive with a bucket of warm soapy water and cloth to make sure that *Rocket* looked its best to impress the visitors. The copper and brass-work was all given a polish with oil and brick dust to make it shine brilliantly. Occasionally more fuel was put on to the fire as the whole boiler heated up. The water level in the gauge glass had risen significantly as the water in the boiler expanded with the heat, and the boiler water level in the glass became slightly livelier as the water started to circulate within the boiler barrel and firebox. Also, there was evidence of a small amount of pressure on the wooden marker float of the mercurial gauge, but this was deceptive, as this slight pressure rise was not caused by steam being generated at this early stage, but by expansion of the air trapped above the water level within the boiler as it warmed up and expanded.

'All the moving parts were given a liberal dose of oil from the can, and beef tallow oil was poured into the tallow pots on the top of each cylinder. These pots were displacement lubricators, which allowed a small amount of steam to condense in the bottom of the pot thus displacing a small amount of oil out into the moving parts of the piston and cylinder.

'The cylinder drain valves at the base of each of the cylinders were opened to drain condensation that might have formed in the steam chests. More coke was added to the firebox to build up the fire evenly over the whole grate. By this time all the wood had burned away and the pressure manometer was showing about a third of full pressure.

'The axle boxes of the locomotive, tender and two carriages were checked for water content by lifting the metal lid and using a metal syringe from the tool kit to suck a small quantity of whatever was at the bottom of the lubrication reservoirs. By vigorously pushing the sample out of the syringe it was easy to see any trapped water spluttering out and eliminate the water. The lubrication reservoirs on all four wheel bearings and the carriages were topped up with fresh lubricant ready for the journey, and the worsted trimmings that fed the oil slowly into the bearings were replaced into their holes, ready to deliver small amounts of oil for the next few hours. By this time, some three hours after lighting the fire, the pressure was showing on the gauge to about half of its maximum, and steam was to be seen fizzing from various flanges, fittings and joints. The boiler water level was up to just above half a glass and the locomotive was nearly ready to make the first move of the day.

'Our driver, Mr George Stephenson, arrived, and after the normal pleasantries he climbed up on to the footplate, first checking the boiler water level on the gauge and the boiler pressure shown on the manometer to see that I had prepared his locomotive correctly.

'After a quick check of the fire and a final wipe round to make sure that the locomotive was ready for visitors, our driver climbed down to welcome the guests who were arriving and being shown into their carriages. It was customary on these occasions, if one of the guests was particularly keen to learn about the locomotive, to invite them to join us on the footplate for the journey. The footplate was kept as clean as possible in anticipation of such an event. It has to be said that the light-coloured finery that some of the ladies were wearing at these times was not entirely suited to the dirty and dusty conditions of Rocket's footplate. In their excitement some seemed oblivious to the dust and dirt, and were more taken up with the charm and wit of Mr Stephenson, with his rich Geordie brogue and charm.

'As departure time approached it was necessary for me, as the fireman, to maintain a balancing act with the fire, water level, and steam pressure, to make sure that the locomotive was ready for the journey without wasting fuel and steam. As the steam pressure approached the safety valve's set limit, the valves started to sizzle, indicating that the locomotive boiler was up to working pressure and about to lift the valves to release excess steam. By opening the regulator slightly to allow a small amount of steam into the cylinders, with the cylinder drain cocks open and the hand reversing gear unlatched it was possible to safely pass some of the surplus steam into the cylinders and steam chests to warm them through ready for action, without risk of the vehicle moving.

'At the appointed hour, Mr Stephenson took his place on the footplate and the guard indicated with his whistle and flag that the train was ready to depart. A few light shovelfuls of coke on to the fire just before departure, spread evenly around, ensured that the fire would be able to cope with the changes about to happen within the boiler as steam was drawn off into the cylinders.

'Mr Stephenson, having acknowledged the guard's flag with a wave, latched the valve gear into the starting position and opened the regulator by bringing the handle slowly and firmly to the centre (full-open) position. Steam could be heard leaving the boiler and entering the steam chests. As the steam entered the relatively cold surfaces of the iron chests and cylinders condensation took place, and water started to pour from the cylinder drain cocks of both cylinders as the incoming steam drove the water out. The train started to move slowly. No sound could be heard from the chimney at this stage due to the noise of the water and steam escaping from the drain cocks.

'The train, consisting of Rocket, its tender and two passenger coaches with their excited passengers, slowly started to accelerate. After several driving wheel revolutions Mr Stephenson used the rake to lean forward and shut the cylinder drain cocks. The loud noise of escaping steam ceased and Rocket started to move forward quietly and purposefully. Then it was possible to start hearing the noise of the exhaust in the chimney, as each spent quantity of steam escaped into the chimney with the familiar deep chuff associated with the exhausting steam being fired upwards. This starting procedure had a profound effect on both the steam pressure and water level. The steam pressure fell by about a quarter and the water level dropped considerably. The steam pressure lowered as the stored steam in the boiler was used up to power the train forward. This lowered pressure encouraged more of the stored water in the boiler to boil away into steam. This meant that the boiler needed more water, as well as having to make more steam. So began my job as a fireman that day, with the fireman's great dilemma and challenge of matching water, fire and steam pressure needs ... Turning the crosshead pump on fully to restore the water level would lower the pressure even more, as cold fresh water was pumped into the boiler, so the first thing was to get more coke on to the fire. I thickened the fire to produce more heat with several shovelfuls of coke. I closed the firebox door between each shovelful to stop cold air being drawn into the fire and cooling the boiler even more. As this fuel started burning, to produce results in the form of rising steam pressure, I turned the crosshead pump partially on to reinstate the water that had been lost from the boiler during the starting process.

'By this time the locomotive and train were heading away at a fast walking pace from the Liverpool Edge Hill terminus through Olive Mount cutting, with its steep sandstone side. Navvies could be seen working away at the steep sides of the rock faces, securing or freeing loose and dangerous rock to prevent rock falls. The regular rhythm of the several sets of wheels passing over the rail joints echoed back from the cutting side. This gave a useful indication of the train's speed.

'The steam pressure was now starting to climb back up to the design value of 50 pounds per square inch as the fuel burned through. The fire was showing through the occasionally opened door as white hot and incandescent. The exhaust blast from the spent steam drew in through the fire-bars fresh air for combustion, making the fire look like a large blacksmith's forge.

'Mr Stephenson was able to move the regulator to a partially open position to steady the acceleration to a constant speed. This reduction in steam consumption meant that the crosshead water pump could be put fully on, to get the water level back up to the starting condition. More coke was put on the fire in small quantities but often, and both the steam pressure and water levels could be seen climbing slowly back up to their original values as the locomotive recovered from the torment of starting with a cold engine! Mr Stephenson was able to acknowledge proudly the waves of the wide-eyed bystanders who had turned out in the hope of seeing this newfangled marvel of steam locomotion go by.

'The passengers were delighting in the excitement of travelling through the outskirts of Edge Hill and the countryside faster than they would if they were on horseback. The speed settled to a steady fifteen miles an hour and more coke and water were added to *Rocket*'s boiler to sustain the steam production. *Rocket*'s exhaust beat, heard so distinctly during the acceleration period, had settled back to a gentle, rapid chatter.

'In a matter of only a few minutes of proceeding like this we faced the steep climb up Whiston Plane. This was a stiff, one-and-a-half-mile climb at a steady gradient of 1 in 96 up to the Rainhill Levels, where *Rocket* had acquitted itself so brilliantly in the locomotive trials some few weeks earlier. The directors of the Liverpool and Manchester Company had expected that they would be installing steam winding engines to pull the loaded trains from Liverpool up on to the Rainhill Levels; but Mr Stephenson had persuaded them to keep their options open to the possibility of steam locomotive haulage throughout the route. In this he had been proved right.

'A few minutes before the foot of the incline was approached, I stoked the fire with several more rounds of the shovel. Now the whole fire was much deeper. As the engine and train started to climb, Mr Stephenson opened the regulator to the full open position and the exhaust beat sharpened in the chimney. Doing

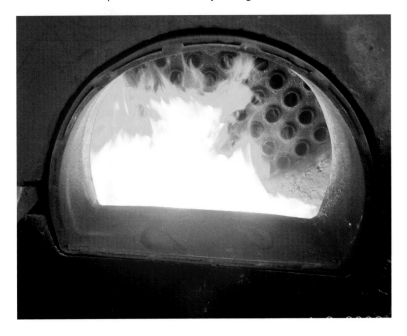

BELOW **The fire in the 1979** *Rocket* **replica whilst at speed during the** *Timewatch* **Trials.**

the fireman's job was always made more difficult going up hills, as the water tended to run towards the back of the boiler, and the gauge glass showed an artificially high level. With this in mind, I had to keep the pump running to ensure that the water level was kept temporarily higher than normal.

'The engine slowed somewhat with its heavy load, and the bark from the chimney became a regular hoarse rasping sound as *Rocket* settled into its own rhythm and took on the steady climb. Mr Stephenson must have allowed himself a few moments of private pride and pleasure during the ascent of the incline (knowing in his heart that the once planned stationary steam engines would not now be needed on this railway). As his machine breasted the summit the regulator was adjusted to level running, and we slowed down for the end of the demonstration run. I kept the firebox door slightly open to admit some cool air to the firebox, as the fire was still white hot from the exhaust blast generated by *Rocket*'s exertions to climb on to the level at Rainhill, where our journey along the new, partially completed L&MR was to end.

'An admiring crowd had gathered at the Rainhill site and they cheered their approbation to join with that of the passengers in the carriages. The carriage hand brakes were applied and the engine was uncoupled and ran forward to be turned round ready for our return to Liverpool, and by means of the points system the engine was able to be attached to the other end of the carriages ready for departure.

'Several of the passengers came up to admire *Rocket* and to ask questions about how it had managed to perform so well on the journey from Liverpool and to engage with George Stephenson to ask him questions. Nobody thought to ask me any questions.

'Once they had their answers the guard ushered them back up the steps into their carriages, and when all was ready everyone waved farewell to the Rainhill onlookers. The train set off back whence it had come, to the top of the incline and back to Liverpool. The guard had been instructed to use his brake to keep the speed of the train under control and reasonably slow on the down gradient. The fire needed much less attention on the way down

the Whiston Plane, as the resistance of the train was balancing the tendency to run away down the hill. By keeping the regulator shut there was a certain amount of engine braking from within the cylinders, resisting the tendency for the train to run away. The passengers arrived back at Liverpool Crown Street station exhilarated and impressed with what they had witnessed.

'There was much adulation for *Rocket* and Mr Stephenson as the passengers left for home. When this was over, the locomotive was then turned again, and I thoroughly serviced, cleaned and lubricated it in preparation for its next demonstration run later in the day.'

The real significance of the capacity of a steam locomotive to haul a train up hills was possibly wasted on the passengers, who were just enjoying the exhilaration of the ride. But the directors of the L&MR must have been rubbing their hands together at the prospect of *Rocket* and train making the full journey from Liverpool to Manchester without the disruption of having to use stationary engine-powered rope haulage to get up and down the gradients along the route. Although the trains would have to stop briefly at the intermediate stations along the route for commercial purposes, and for taking on water, the journey time for the full trip was potentially so much shorter than had been originally planned. In comparison to the equivalent journey by stagecoach the time saved must have seemed miraculous.

Edward Entwhistle's account

The last contemporary account comes from 'The Rocket and the world's first Railway from the narrative of Mr Edward Entwhistle', as told to Walter Wood of *The Royal Magazine* in 1909 under the series heading of 'Survivors' Tales of Great Events'.

'This unique story takes us back to the remote days when travel was revolutionised completely by the opening of the Liverpool and Manchester Railway in September 1830. The *Rocket*, most famous of all the strange locomotives that were originally brought out, was the invention of George Stephenson and is still preserved in the

South Kensington Museum as a national relic. Closely associated with Mr Stephenson was Mr Edward Entwhistle, a native of Lancashire, who on many journeys drove the *Rocket*. Seventy years ago Mr Entwhistle emigrated to America where he has lived ever since. More than half a century of his life has been spent in the city of Des Moines, the capital of Iowa, and there he remains, in the enjoyment of excellent health and a marvellously good memory.

'"I was born on March 24th 1815 – in the year of Waterloo. I had not much education to speak of over and above the practical teaching of glorious men like George Stephenson; but in my young days lads learned a lot verbally, and I was brought up on all sorts of tales about Napoleon and the Iron Duke. I believed like everyone else that Wellington was a man of iron; yet when I saw him, for the first and only time in my life, it was under conditions which showed how even a great soldier like him can be upset by suffering and death, when both are unexpected and come upon a man in time of peace.

ABOVE Conjectural image of *Rocket* in un-rebuilt form arriving with a passenger train through the Moorish Arch at Crown Street station, Liverpool. It gives a feel of what would have greeted Fanny Kemble's adventure with George Stephenson. *(Rainhill Railway and Heritage Society)*

'"When I was fourteen years old I was apprenticed to the Duke of Bridgewater. For about two years I was working in his machine shop and engine room in Manchester. I soon picked up pretty well all there was to learn about practical engineering and of course like everybody else was aware of the existence of *Rocket* and other locomotives which were then being brought into being by inventors. Men laughed and scoffed at the idea that a locomotive could be built which would travel with passengers and draw carriages faster than horses could run.

'"In those days Manchester and Liverpool were connected by canal and turnpike, and as they were not the times of rush and hurry, travelling even by stage coach was very slow

ABOVE A train negotiating Olive Mount cutting in 1831 as work to complete the sides of the cutting was still in progress, illustrating some of the many dangers of early rail travel.

errand, and that was to find a man to drive the *Rocket* that was to take part in the first run between Manchester and Liverpool at the opening of the line.

'"Now there were not many men who had the same faith in the locomotive that the inventor had; and although he spoke in the shop for a long time, he could not persuade anybody to volunteer. By and by Stephenson turned away disappointed and I dare say disgusted. Well when Stephenson was leaving the shop the foreman went after him and said 'I don't know any man that I can recommend but if you can get leave from the Duke's steward to take that lad there as an engineer, I'll warrant him.' Stephenson came back and looked at me and evidently his inspection was satisfactory, because he went and got leave from the steward, and on the Saturday afternoon I entered the shop where the *Rocket* was in Manchester.

'"I managed the *Rocket* all right and got to know every joint about her. We spent that Saturday in getting the *Rocket* ready and on the Sunday we got steam up, and George Stephenson and myself took her out of the shed for a trial run before the big trip which was to be made on the Monday when the Railway was to be opened.

'"When we were actually driving the *Rocket* we got a pressure of from 60 to 80 pounds of steam per square inch, as far as I remember and once we raised it to nearly 100 ... I went with Stephenson to Liverpool on the *Rocket*. I was to do the firing and driving and Stephenson was to stand by me all the time. I knew what to do, how to stop and start the *Rocket* and feed the boiler, but I was very young, and perhaps nervous. We had just enough room to stand on and no more, and the good old engine jumped and jolted in a way that would scare a driver today. There was no spring in the sleepers and there was little or none in the *Rocket* herself so that when I wasn't looking after the firebox or the boiler, I was busy holding on to keep myself from being thrown on to the line. We went at what was looked upon as a marvellous speed, and to a young man like me, accustomed to seeing the leisurely method of travel by canal and turnpike, the *Rocket* seemed to go at a terrific rate. She had done nearly twenty miles an hour at one time but the average was much less.'"

and it took as long to cover the thirty-six miles or so of road as it takes nowadays to go between London and Newcastle.

'"A railway was being made between the two Lancashire cities and Stephenson was the engineer who superintended its construction. Only a man like Stephenson was able to carry the scheme through because the most extraordinary hatred was shown towards the idea of a railway that would draw passengers. Even when the railway was finished, and the trial trips were ready for running, Stephenson found it hard to get drivers because it was believed that the engines would blow up and kill everybody.

'"One day Stephenson came into the shop where I was working. He was on a very strange

Edward Entwhistle then goes on to relate the events of the opening day described in chapter one, including the horrific accident at Parkside to Mr Huskisson, and how he stayed with *Rocket* for the next two years. He ends with:

'"For two years I remained with the *Rocket* working her and getting to understand her in every part – and to love her, just as I love her still and always shall. Then for some reason I cannot explain, I became nervous and afraid of myself, young though I was. When a thing like that happens it is time to sever partnership with engine-driving, and I did so."'

As demonstration of the difficulty of accurate recall of past events, well known in police enquiries, Entwhistle's false memory of the boiler pressure of *Rocket* is alarming and cannot be true, because of what we know about the boiler design. The value of writing reminiscences down as soon after the event as possible, as Fanny Kemble did, is thus well shown.

An account of driving a *Rocket* replica

We can contrast the accounts above with one from a driver of the replica that performed regularly at the National Railway Museum. The writer is the museum's former Education Officer, David Mosley, who was a regular volunteer driver of the replica *Rocket* and knew the machine's foibles only too well:

'"Is it difficult to drive?" is a frequently asked question whenever the replica *Rocket* is operating. "Come and have a try" is a reply which usually delights the enquirer and in truth, no the replica isn't so difficult to drive but oh so idiosyncratic!

'A gentle pull on the regulator, a reassuring hiss of steam down the steam pipes – and away she goes – sometimes! For remember *Rocket* has a fixed cut-off and the valves may have come to rest in a position which doesn't allow enough steam pressure to come into play to enable the driver to move off as he or she would wish. So the valves have to be adjusted by hand to bring the appropriate forces to apply. This "striking off"

becomes easier with experience, just a matter of observing the position of the cranks and adjusting the valves accordingly.

'It wasn't always so straightforward, the writer's "record" for persuading *Rocket* to go from reverse to forward gear is an embarrassing twenty minutes, once in the National Railway Museum's Yard with the then Chairman of British Gas on board, and a second time at the Midland Railway Centre attempting to start from Ironville and return to Swanwick Junction!

'Once on the move the next critical consideration is stopping – safely and in the right place – and although the replica *Rocket* is fitted with an up to date, efficient air brake system, regular drivers always make it a matter of pride to stop in the traditional way.

'This is to reverse the engine and use the pressure of the steam to act against the piston rather than with it. Simple, the operation of a foot pedal on the footplate pushes a rod that, in its turn, moves two plates across the driving axle. As appropriate one of these will engage a fixed "tooth" which positions the valves and the locomotive will then be in the gear desired by the driver. He or she must be careful to watch the valve levers on the footplate for "striking off" you recall. They start to waggle so confirming that the locomotive will now go the way the driver expects on steam being applied. And that foot pedal, a source of trial for the newcomer, a simple "heeling a ball" motion to go from back to forward gear, to go the other way a delicate flick of the foot, down and away to ensure that the pedal is firmly located in its position on the footplate, is an acquired art!

'Of course it all depends on steam and the new copper firebox and boiler with fewer larger tubes is a much more predictable unit than the original boiler fitted to the replica in 1979. Keeping pressure between 42 and 45 psi all day – she blows off at 50 – whilst *Rocket* stars in museum theatricals is a great discipline for a fireman!

'Over the years it has been the greatest privilege to be involved with the replica *Rocket* – just think, add brakes and a super-heater and you're virtually at "Evening Star" – what an achievement back in 1829!'

(David Mosley, *Rocket* footplate-man 1982 to 2014)

Dr John Glithero and Dr Michael Bailey communicating through holes in the boiler of *Rocket* during their investigation.

CHAPTER FOUR

Maintaining *Rocket*

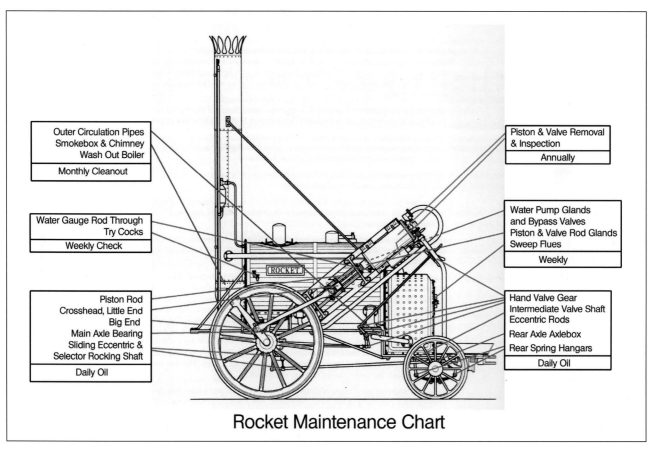

Outer Circulation Pipes Smokebox & Chimney Wash Out Boiler
Monthly Cleanout

Water Gauge Rod Through Try Cocks
Weekly Check

Piston Rod Crosshead, Little End Big End Main Axle Bearing Sliding Eccentric & Selector Rocking Shaft
Daily Oil

Piston & Valve Removal & Inspection
Annually

Water Pump Glands and Bypass Valves Piston & Valve Rod Glands Sweep Flues
Weekly

Hand Valve Gear Intermediate Valve Shaft Eccentric Rods Rear Axle Axlebox Rear Spring Hangars
Daily Oil

Rocket Maintenance Chart

ABOVE *Rocket* **lubrication chart in the style of the Castrol charts popular with post-war motorists.** *(Adrian Lucey)*

Scheduled maintenance programme

In this chapter servicing and maintenance requirements to keep *Rocket* in good working condition are discussed and analysed. Unlike a car, for which servicing periods and requirements are listed in a service handbook without explanation of why they are necessary, the unique nature of *Rocket* offers an excellent opportunity to discuss the rationale behind the maintenance schedule. The frequency of each servicing requirement is suggested, but without contemporary documentation the periodicity is based on conjecture from the author's intimate knowledge of the operation of the 1979 replica *Rocket*.

A steam locomotive was (and still is) the antithesis of a modern car, steam iron or washing machine, which can run for long periods without much knowledge, care or attention on the part of the operator. The design principles on which modern 'sealed-for-life-then-throw-away' machines are based is that they must not demand regular time-consuming attention from the user!

LEFT *Rocket* **replica boiler inspections are made through a disguised manhole in the front tube-plate. Here the annual statutory boiler inspection is under way.** *(Author)*

Rocket, like the hundreds of thousands of steam locomotives that were to follow over the years, was a basic machine which would only have given of its best when understood, nurtured, maintained and properly cared for. The performance of *Rocket* therefore depended on systematic preparation, regular maintenance and disposal within fixed periods. The day-to-day reliability of the locomotive was determined by careful and thorough inspection of every accessible part before each day's running. The inspection would have involved checking for loose fastenings and missing split pins, as well as keeping an eye open for metal chafing, debris or escaping steam, oil or water from the wrong places. This procedure was known as the 'Fitness to Run' examination, and was only to be carried out by experienced, qualified persons.

Dr Michael Bailey and Dr John Glithero established that *Rocket* was involved in a series of mishaps or accidents during its early months and years of use. The Stephensons took these opportunities, whilst it was in the workshops for repair, to update their machine to match the latest thinking as their working knowledge of the locomotive in revenue-earning service developed. The maintenance that *Rocket* needed would have changed as the design evolved throughout the cumulative rebuilding processes.

Below are listed and discussed the scheduled maintenance tasks needed to keep *Rocket* in tip-top condition. A chart in the style of the old 'Castrol Motor Car Lubrication Chart' is included at the top of page 84 to indicate the maintenance and cleaning regime of a *Rocket*-type locomotive. Of course, it is all slightly tongue in cheek, as the knowledge is based on running the replica locomotive using modern oils. It is entirely possible that the original locomotive would have needed different frequencies of attention.

The role of lubrication by means of applying oil to moving parts is taken for granted nowadays, but when *Rocket* was new there was no such thing as motor oil or steam oil. Mineral oil is a 20th-century product. In the absence of mineral oil, animal and vegetable derivatives provided the lubricants that were used in the Stephensons' time. Early ships' engineers were advised to visit the kitchens regularly to collect the fats that ran off cooked meat for use on the

bearings of 19th-century ships' engines. Early railway waggons were lubricated with tallow derived from rendered beef or mutton. Tallow had the advantage that it was fairly solid at room temperature, but as the temperature of the bearing rose the lubricant melted to provide more oil where it was needed. Olive oil also provided excellent properties for lubricating the internal sliding surfaces of steam engines. Even in modern times, when our lubricants come to us in shiny five-litre cans, the preferred oil for railway steam locomotive lubrication is called 'compounded bearing oil' and contains at least 10% rapeseed oil to ensure tenacity of the oil film once movement has ceased.

Water gauge maintenance

The importance of the *Rocket*'s crew knowing accurately what the actual water level is in the boiler has been discussed already in chapters two and three. This level was affected by the locomotive's attitude on the track as the engine moved around, but it was

ABOVE 1979 *Rocket* replica undergoing maintenance in the National Railway Museum workshops. Note 60800, 49395 and 46229.
(Pastel by Roy Wilson)

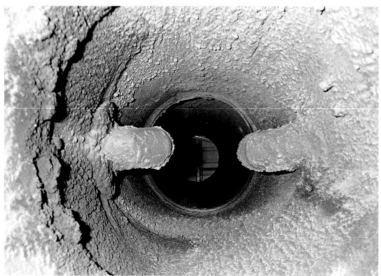

also affected by the condition of the inside of the water gauge glass assembly. It was essential that the water gauge passages were checked regularly as being completely clear of obstruction. (There were removable plugs through which wire could be poked to check this.) Also, the glass tube in the gauge was replaced annually, as the parts of the glass that were immersed in the pressurised, boiling water became partially dissolved over a long period and gradually wasted away. This wastage was, of course, invisible until it was too late.

Soot and ash deposits

Because *Rocket* worked by burning a coal-based fuel (coke) in a confined space, it was inevitable that dirt, dust, ash and clinker would accumulate within the combustion spaces. This debris was deposited within the fire-tubes, smokebox and chimney areas, lessening the ability of the water inside the boiler to accept heat from the hot flames and flue gases. This debris coated the inside surfaces of the firebox and fire-tubes, effectively insulating the passage of heat to the water to make steam. Consequently these surfaces needed to be cleaned back to their original condition regularly to ensure optimum performance.

A long flue brush was inserted through the fire-hole or smokebox to scrub the inside surfaces of the tubes and firebox. The small smokebox at the base of the chimney of the Rainhill-condition replica was cleaned by opening the pair of small doors in the base, which gave access to the interior of the space under the chimney at the front of the tubes. To ensure tip-top performance this would be done after every full day's work.

Water quality and the effectiveness of the boiler

Steam boilers, like domestic kettles, need to be kept supplied with fresh water regularly. The quality of the water in the boiler is critical. Every beverage produced from a kettle has the boiling water decanted into the drink along with slight impurities in the kettle's water. However, the steam generated from that same water quality in a steam boiler is *pure* steam,

and contains none of the impurities that remain within the liquid left in the boiler when steam is drawn off. The amount of impurity in fresh water varies greatly around the country, but hardly anywhere has impurity-free water. So as fresh water is pumped into the boiler of the steam locomotive throughout a day's work, the debris remaining from those impurities becomes more and more concentrated within the boiler space. It no longer has the same properties as clean water and must be drained away and disposed of before another load of clean water is pumped into the boiler.

It may take several days of operating for an unacceptably high concentration of debris to be reached, but the dirty water must be dealt with by draining, washing out and refilling the boiler. One way of doing this is called 'blowing-down'. Once the fire has been raked out and is no longer producing heat, it is possible to use the remaining steam pressure in the boiler to purge the remaining water through a drain valve at the bottom of the boiler barrel. In *Rocket*'s case this did not get rid of the water remaining around the lower parts of the firebox, so further draining of this region was necessary once the pressure had dropped away. This meant taking the locomotive out of service and allowing it to cool thoroughly.

There is a relevant contemporary quote from John Rastrick's notebook written at the Rainhill Trials. The extract shows how well they understood the problems of maintaining *Rocket* properly in those early days. He writes about Stephenson's copper firebox on *Rocket*:

'...the principal objection to this method of construction is the difficulty of getting at the inside to clean it out. The sediment from the Water will mostly settle to the Bottom into the narrow part where it is riveted together, but the

RIGHT Flue brush inside the tubes of the 2010 replica at the National Railway Museum. *(Author)*

scurf or calcareous encrustation will affix itself to all parts of the Inside of the Boiler. It is proposed to clear out the sediment by making a small Hole on each side towards the bottom to get the hand in, but as the distance between the Sides of the Boiler is only three Inches, and as there are many distance pieces in the way, the withdrawing of the sediment will be a tedious and difficult operation and the scurf cannot be effectually scaled off from the inside of the Boiler without taking it to pieces. How often this may be required to be done must depend entirely on the nature of the Water with which the Boiler is fed and can only be learned from experience.' (John Rastrick's notebook of 1829, Science Museum Group collection.)

Corrosion

*R*ocket's boiler barrel was made out of wrought iron, and corrosion over time of the ferrous material was always an issue. Iron rusts when exposed to water over long periods. Consequently washing out of the boiler was always followed by a thorough inspection of its internal surfaces using a paraffin-soaked lighted flare. The red colour of the water during the washout was always an alarming indicator of how much of the boiler's interior had been dissolved during the latest period of operation. The detailed inspection therefore involved examining surfaces and components generally below the water level that might have

rusted excessively beyond *normal* oxidation. Nowadays this detailed inspection for corrosion would be scheduled to take place every ten to twenty full days of steaming. No doubt the L&MR would have had their own regular inspection arrangements based on experience with early steam railways.

In *Rocket*'s boiler, solid deposits built up and choked the external copper water-circulating pipes as well as the lower regions of the copper firebox saddle. During boiler washouts and subsequent inspection, this was a particular challenge. The water circulation speed is so slow and gentle that it cannot be relied upon to scour out deposited solids, so occasionally the external pipes had to be removed and the sludge purged from them. On such occasions those corresponding pipes in the replica were found to be more than 50% choked by debris. A tannin-based water treatment applied to the feed water helped to reduce the corrosive effect of fresh water entering the boiler.

An important aspect of the maintenance of *Rocket*'s boiler in service was to make sure that internal surfaces in contact with water were kept in good condition. That applied to the surfaces of the fire-tubes especially. Scale and oxidation on these surfaces impaired the ability of the heat to transfer efficiently through to the surrounding water. To that end *Rocket* had a so-called 'manhole' between the chimney and the second safety valve on the front of the boiler at the top. This was held on a ring of bolts that, once removed, allowed a small person access to the inside of the boiler barrel above the tubes for inspection and cleaning purposes. This 'manhole' had a spring-loaded safety valve attached to it originally.

Lubrication

All the moving parts of *Rocket* required lubrication both internally and externally to allow them to run smoothly. The system used

BELOW View inside 1979 replica boiler, from the front. The water level and the scaling on the tubes are evident.

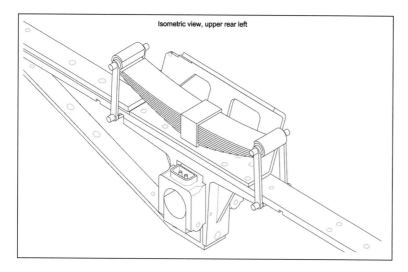

Isometric view, upper rear left

LEFT Isometric drawing of *Rocket*'s front left-hand axle box arrangement. *(John Glithero)*

for lubrication of a locomotive was based on the custom and practices of early textile and mill machinery maintenance. A small reservoir pot containing lubricating oil was positioned above the site where the bearing assembly was operating. A thin tube led down to the bearing from within the pot and the top end of the tube was positioned above the liquid lubricant level.

Then a 'trimming', made of worsted wool wrapped in copper wire, was inserted into the top of the tube, and the tails of the trimming were allowed to droop into the surrounding lubricant. This oil then wicked up into the worsted wool trimming and, because the lower end of the trimming was higher than the top end, slowly siphoned out of the reservoir into the bearing one drip at a time. By having multiple strands to the trimming the quantity of oil delivered to the bearing could be increased pro rata. What is fascinating is that such trimmings had to be made from worsted wool, which has longer, smoother fibres than ordinary wool. It was this that enabled them to act as wicks to deliver oil reliably to bearings while a machine was running.

When the day's work was finished the trimmings were removed from their holes and left in the liquid reservoir, ready for inserting back into the feed tubes the next time the locomotive operated. The manufacture of worsted wool for making trimmings is one of the many craft knowledge skills that engineers running heritage machinery must acquire and pass on.

Bearing adjustment

Modern machinery is often relegated to the scrap heap as soon as a breakdown occurs. Maintenance and repair consists largely of substitution of new parts to solve any simple problems identified. To understand the maintenance and repair of *Rocket* we have to understand that the men who designed and built it lived in a period when it was believed that every moving part could be adjusted and fitted to keep a machine in perfect working

RIGHT Worsted wool trimming of the type used on *Rocket* and all subsequent steam locomotives to deliver oil to bearings. *(Author)*

BELOW Oiling the 2010 replica, showing soaked trimmings inserted into lubrication holes to the journal bearing, with 'tails' in the oil. *(Author)*

order. This was partly an economic imperative. Machines were too costly to be scrapped. It is hard for us to imagine such a world, when we see modern machines such as washing machines, computers and cars scrapped when they break down or wear out.

An examination of *Rocket*'s crosshead and connecting rod components is a good example of the maintenance philosophy referred to above. The selection of materials was vital to ensure the long and trouble-free life of moving parts. Modern engineers do not select materials for the manufacture of components that work inappropriately against each other. However, the concept of 'friction pairs' was well known as a result of empirical experiments long before *Rocket* and the modern study of what is known as tribology. It was proven that identical engineering materials running together were generally poor performers as friction pairs, whereas bronze or white metal running against steel or cast iron gave much better performances over time. Given correct fitting and lubrication such a friction pair would continue to give satisfactory service for many hours.

Eventually, however, clearances between one component and its mate would increase to the point where adjustment would become necessary in order to re-establish the correct running clearances between the two moving components. This work was carried out by locomotive fitters. Their job was to separate the two halves of the bearing and, by careful filing, remove metal from the clamping surfaces where necessary until the original manufacturer's clearances were re-established. This enabled readjusted bearings to be put back in service very quickly for another spell of duty.

Glands on sliding joints

The shafts that slid in and out of *Rocket*'s cylinder mechanism were sealed with what were called 'stuffing boxes' or 'glands'. With steam pressure on one side and atmospheric pressure on the other, some form of sliding seal against leakage was necessary. This seal was achieved by placing rings of woven, graphite-lubricated yarn into the cavities and squeezing them longitudinally with the gland nut. The yarn

ABOVE Today, when we throw away machines with small faults, it is hard to imagine an age when parts like those of *Rocket* were infinitely repairable.

LEFT Connecting rod and crank pin assembly of 1935 replica, showing how it strips down to allow parts to be carefully adjusted by hand to re-establish a perfect fit. *(Author)*

LEFT Little end and crosshead assembly of 1935 replica, showing how adjustment for wear is catered for in manufacture. *(Author)*

ABOVE Valve-rod gland to seal valve rod into steam chest. The hexagon nut traps and squeezes the yarn to seal on to the shaft. *(Author)*

ABOVE Pump rod gland assembly. The hexagon nut tightens the graphited yarn on to the shaft. *(Author)*

LEFT Piston-rod gland on 1935 replica to seal off lower end of cylinder as piston rod slides in and out. *(Author)*

BELOW Gland packing removal tools – the corkscrew is wound into the remains of the old gland material, which is pulled out and disposed of properly. *(Dai Jones and Steve Thorpe, Talyllyn Railway)*

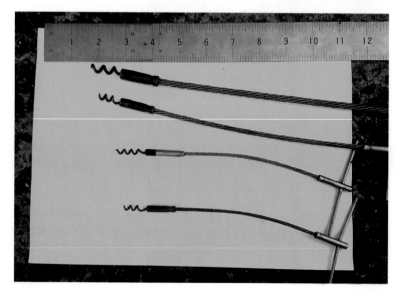

was trapped in the cavity, and the only way it could move when the gland nut was tightened was to squeeze the sliding shaft more tightly. The yarn lubricated itself on the sliding shaft during movement but prevented the escape of steam.

There were five stuffing boxes in *Rocket*'s design. One was positioned on each of the main power piston rods at the front of each of the cylinders. Another was on each of the valve spindles where the rocking valve shaft passed the motion to the oscillating valve spindle into the valve chest. Finally, one was on the sliding joint at the lower end of the crosshead water pump, sealing the pump ram as it passed in and out of the pump body.

All stuffing boxes eventually develop leaks between the yarn and the shaft if they are not regularly adjusted. In *Rocket*'s design, hexagonal flats on the body of the brass gland nut allowed the stuffing box to be tightened regularly. Over-tightening caused resistance to the sliding motion, which was undesirable. Knowing how tight to spanner up the gland nuts is an acquired technique and part of the skill of a good locomotive fitter. The stuffing boxes on *Rocket* would probably have been adjusted a small amount (less than half a flat on the hexagon nut) before each day's work. At some point after many adjustments the nut would have reached the bottom of the threaded hole and could not be tightened further. When that stage was reached the gland nut was unscrewed and pulled clear of the cylinder. Then a tool similar to the ones shown at the foot of page 92 was used to extract the remnants of the old gland packing so that it could be replaced by new graphited yarn.

This maintenance operation would probably have been done annually.

Grate

When coke burned vigorously on the fire-bars within a firebox and then demand suddenly ceased, there was a tendency for the combustion process to generate clinker. This was especially so if the fire had been partially shut down and then restarted. The clinker, which is an ugly crystalline form of ash, tended to fuse together with the fire-bars on which it was generated.

To maintain *Rocket*'s grate after each run and after the fire had cooled down, all traces of clinker from the fire-bars were removed so that the grate was completely free ready for combustion and the next steaming.

The length of the fire-bars was critical and was carefully controlled. The fire-bars rested on horizontal supports at the front and back of the firebox base. They needed longitudinal clearance of at least one-eighth of an inch when cold. If this clearance was not maintained when fire-bars were renewed or replaced, there was a tendency for the fire-bar to expand until it locked within the firebox space, and then distorted sideways permanently (the only way it could move) when it got red hot under a working fire. This made gaps and distortions between the bars that were unacceptable for the fuel to rest and burn on.

Tender water barrel

Traditional wooden barrels only held water reliably when they were thoroughly wet and all the staves were swollen. In order to satisfy this condition the water barrel was always filled up at the end of a day's operation. Obviously, if

ABOVE The fire grate from the 1979 replica set out on the floor of the workshop.

LEFT Clinker sample removed from a locomotive grate. (Author)

BELOW 1979 replica at the National Railway Museum launch of the 'Virgin Super Voyager', showing the sherry-barrel origins of the water supply in the tender.

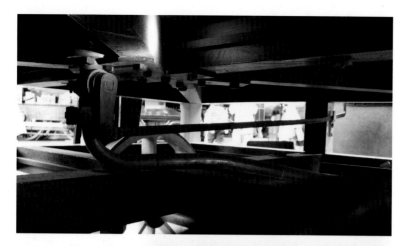

ABOVE 1935 replica tender water shut-off valve. *(Author)*

RIGHT Good flange condition of 1935 replica *Rocket*. *(Author)*

RIGHT Worn flange on locomotive driving wheel showing sharp tip, which can risk splitting poorly fitted point blades. *(Author)*

the locomotive was to be out of use for a while then the barrel was emptied. However, once the barrel was drained leakage was inevitable until all the wood had become thoroughly soaked again and the barrel stave joints had swelled tight once more.

A valve was provided on the underneath of the barrel to turn off the water supply to the crosshead water pump. This valve had to be turned off after each day's running. If it was not, as the steam in the boiler cooled it formed a vacuum that sucked the water from the tender barrel into the boiler, and an unwary crew would be caught out the next time the engine was steamed when they found that the boiler was full to the very top!

Wheel wear

Metal wheels running on metal rails always show signs of wear over time, especially when the track curves. Part of the maintenance regime for *Rocket* would have been to keep a check on the state of the rolling surfaces of the wheel tread and flange.

Nowadays there are measured standards of wear for rolling stock that must not be exceeded, and there are tyre wear gauges that engineers can use to check how much wear the tyre has sustained. But in *Rocket*'s day there would be inspections when the locomotive came into the works for servicing, when wheel surfaces would be restored to new condition as necessary. This would be achieved by replacing the tyre or turning the wheels in a lathe so that material on the outer surface was machined away to restore the profile and ensure that it was the same diameter as the wheel on the other end of the axle. Tyres were thick enough to allow this process to happen several times before the old tyre was scrapped and replaced with a new full-thickness one. With the largely straight track of the Liverpool and Manchester Railway, and the minimal use of *Rocket* after the opening of the line, it is unlikely that the tyres would have needed replacing due to wear. The wooden wheel centres were much more vulnerable to wear and tear, and there is evidence from the archaeological 'dig' on *Rocket* that the wheels are not original.

RIGHT Loose wooden spokes on 1935 replica wheel. *(Author)*

Bailey and Glithero found that the left-hand wheel of the existing *Rocket* showed more wear than the right-hand wheel. They stated that locomotives of that period with wooden wheels were likely to have been supplied with spare wheels and axles to allow for the punishing service regime of working locomotives.

A wooden wheel on a horse-drawn cart can be used even when the spokes, rim and hub components have come slightly loose. A common cure for loose components was standing wooden carts in the duck pond! Such a solution was not available to the fitters maintaining *Rocket*'s wheels.

If the wheels were not the right distance apart the vehicle would derail, so careful and thorough driving wheel inspection was no doubt a daily task for those maintaining *Rocket* during its working days.

Under worn conditions the gently tapered flat wheel profile with a rounded flange can alter drastically. After heavy wear the flange can become thin and quite sharp. This can have a disastrous effect on the way that the wheels negotiate a pair of points, where a thinned blade of steel rests against the running rail. The danger is that the thinned flange, on approaching the blade lying close to the paired rails, splits the thinned blade from the main rail and the train is derailed.

Conclusion

This overview of maintenance for *Rocket* has focussed on the routine jobs that anyone owning *Rocket* or one of the working replicas would have to carry out to run a reliable locomotive. But there will always be some out-of-the-ordinary events in the life of any steam locomotive. One in particular concerning the 1979 replica *Rocket* comes to mind, which is worth recording just for the sheer unusual nature of such occurrences, their consequences and the lessons we learned.

This event concerned the National Railway Museum's workshop supervisor Dave Burrows

ABOVE Poorly maintained point-work which would allow a worn flange to pass through the gap on the left and derail the train. *(Author)*

and a loose ash pan. It occurred during a routine operating session with the public in the National Railway Museum yard, when the front fastenings that hold the ash pan to the underside of the firebox became detached. The ash pan (which of course was not fitted to the real *Rocket*) swung downward at the front, with the back still firmly held to the rear of the firebox, striking the track whilst the locomotive and train were travelling forwards. The force of the impact leapfrogged the back of the locomotive into the air and back down again, and the rear wheels returned not to the rails but to the sleepers, causing a few tricky moments for the crew. Happily the engine was re-railed and the crisis was averted. However carefully engineers devise and plan a maintenance regime for a machine, they must always be ready for unexpected twists and turns in the day-to-day operating of machinery.

CHAPTER FIVE

Rocket
replicas

1935 *Rocket* replica
at the National
Railway Museum.
(SSPL)

The concept of replicas

The desire to make a representation of a favourite object or figure is a deeply felt and basic instinct. Human beings have replicated artefacts that are special to them since civilisation began, as diverse as cave paintings of favourite animals, statues of holy figures or model ships in bottles. However, Stephenson's *Rocket* has had more than its fair share of full-size replicas made over the years, with at least ten known about. This chapter examines what is known about these models, with a special emphasis on their significance. All the replicas or reproductions represent *Rocket* in its Rainhill Trials form, although that iconic profile lasted barely a year.

So why have there been so many replicas made of *Rocket* in its Rainhill guise? It is because the shape, colours and form of the Rainhill *Rocket* are visually attractive in the same way as the Model T Ford automobile and Spitfire aircraft, and immediately strike a chord with the public. The table below itemises the known replicas.

The 1881 Crewe replica

The first significant recorded *Rocket* replica, Number 1 in the table, was created by the Crewe Works of the London and North Western Railway Company. This company had absorbed the L&MR into its burgeoning empire in 1846. Their non-working *Rocket* replica was built to take part in various pageants held to celebrate 50 years since the opening of the L&MR.

All known full-size replicas of *Rocket*
(with thanks to Dr Michael Bailey and Dr John Glithero)

Replica number	Year	Manufacturer	Type	For whom made	Present location	Comments
1	1881	Crewe Locomotive Works	Non-working	London & North Western Railway	Scrapped in 1975; parts used for 1979 replica	Rebuilt in 1911
2	1892	Mount Clare Shops, Baltimore, Maryland	Wooden, non-working	Columbian World's Exhibition	B&O Railroad Museum, Baltimore	Initiative of Colonel J.G. Pangborn
3	1923	Buster Keaton/Joseph Schenk, Hollywood, California	Non-steam operable	Film prop for silent movie *Our Hospitality*	Smithsonian Institution, Washington, DC	Movie starred Buster Keaton
4	1929	R. Stephenson & Co Ltd, Darlington, Works No 3992	Operable	Henry Ford	Henry Ford Museum, Dearborn, Michigan	On display
5	1930	Derby Locomotive Works	Non-working	London Midland & Scottish Railway	Probably scrapped in 1930s	Made by 'T. Robinson', Liverpool, for *Rocket*'s centenary
6	1930	R. Stephenson & Co Ltd, Darlington, Works No 4071	Sectioned	Museum of Peaceful Arts, New York	Shawnee Mission, Kansas (Current location unknown)	Privately owned
7	1931	R. Stephenson & Co Ltd, Darlington, Works No 4072	Operable	Museum of Science & Industry, Chicago, Illinois	Museum of Science & Industry, Chicago, Illinois	On display
8	1935	R. Stephenson & Co Ltd, Darlington, Works No 4089	Sectioned	Science Museum, London	National Railway Museum	On display
9	1979	Locomotion Enterprises, Springwell Workshops, under Mike Satow	Operable	National Railway Museum, York	Parts used in 2010 replica creation	
10	2010	The Flour Mill, Bream, Gloucestershire, under Jim Rees & Bill Parker	Operable	National Railway Museum, York	National Railway Museum, York	Operated at York and elsewhere

By this time the original *Rocket* was safe inside the Science Museum in South Kensington, London. There was some debate about what shape the replica version of *Rocket* should be. Robert Stephenson & Company had put down on paper their best guess of what they thought *Rocket* probably looked like at the end of its working life. The Crewe replica was therefore built to this design and no one challenged it. More on this later.

It is understood that replicas were made to try to represent *true* versions of the iconic *Rocket*, but there were opportunities for fakery too. The photograph shown below right, for instance, raises immediate questions. We know that George Stephenson died in 1848, and we know that *Rocket*'s cylinders lasted only a year in the steeply inclined position until 1831. There is also something odd about the shape of the firebox. So it is not George, and it is not *Rocket*! The photograph is a fake. The Science & Society Picture Library (SSPL) caption for this image reads:

'George Stephenson and the *Rocket*
An interesting faked undated photograph purporting to be of George Stephenson and the *Rocket* steam locomotive at around the time of the opening of the Liverpool and Manchester Railway (1830). The fact that this event took place only shortly after the invention of photography (1826) and before Fox Talbot's invention of the negative/positive process (1835–1839) may have made the forgery easier to spot. George Stephenson (1781–1848), nicknamed "the father of railways", was the engineer for the Stockton and Darlington Railway (1825) and the Liverpool and Manchester Railway (1829). His son Robert (1803–1859) was also an engineer and designed the "Rocket" locomotive which won the Rainhill Trials in 1829.'

This example demonstrates how careful historians and researchers must be in *triangulating* evidence to substantiate the veracity of historic photographic images. Note that the caption does not try and explain the anomalies, but even before the advent of Photoshop there were ways of faking reality. We need to know more!

We know from John Rastrick's notebook,

made at the Rainhill Trials, that the top of the original *Rocket*'s firebox was flat, not sloped as is shown in this photograph. We know from Dr Michael Bailey and Dr John Glithero's findings that by 1830 the cylinders of *Rocket* had been rebuilt to the near-horizontal alignment seen in the present Science Museum-based remains when it was approximately a year old.

The locomotive in the picture is believed to be the non-working 1881 Crewe Works replica (Number 1 in the table). It was used later on for the photographs to commemorate the

ABOVE 1881 *Rocket* replica at Crewe Works.

BELOW Fake photograph that purports to be the original *Rocket* with George Stephenson. The issues are discussed in the text.

ABOVE The 1881 *Rocket* replica being used to celebrate the construction of the 5,000th steam locomotive built by the London and North Western Railway at Crewe Works.

ABOVE RIGHT The 5,000th locomotive was completed on 29 April 1911.

5,000th locomotive built at Crewe in 1911 (see above). This replica was somewhat rebuilt to correct the shape after the construction of the Robert Stephenson replicas in the late 1920s established once and for all the external firebox shape. This was replica Number 4 in the table.

So how did the engineers at Crewe in 1881 get the shape of the firebox in replica Number 1 so wrong? Thankfully there is a very logical and convincing explanation, which Bailey and Glithero revealed in their work.

In chapter one, 'The story of *Rocket*', we are reminded that the remains of the original much-modified *Rocket* were sent back to Robert Stephenson & Co with a view to simply replacing all the missing parts so that *Rocket* could be considered for inclusion in the Great Exhibition of 1851. (Bailey and Glithero tell us that the locomotive remained at Robert Stephenson & Co from 1851 to 1862.) The conclusion was that too many missing parts at that time meant that there was no chance of it being ready in time to exhibit, so the idea was abandoned.

Some years elapsed before the chief draughtsman at Robert Stephenson & Co, a Mr J.D. Wardale, prepared drawings of *Rocket* measured from the remains that were stored incomplete (probably in the open air) 'standing in a lumber yard at South Street'. We know that many of its high-value, non-ferrous parts, like the copper firebox, were missing before *Rocket* came to Newcastle. We can perhaps imagine the would-be creators of the 1881 *Rocket* replica examining the remains and, in discussion with the Robert Stephenson & Co staff at South Street, coming to the conclusion that the firebox top must have had the slope that was clearly shown in J.D. Wardale's drawing at the top of page 36. Perhaps their thinking was influenced by the pronounced 45° chamfer shown on some early depictions of *Rocket*'s firebox. Even the penny-in-the-slot animated model of *Rocket* shown on the left, used for entertainment on railway stations, has the sloping firebox top.

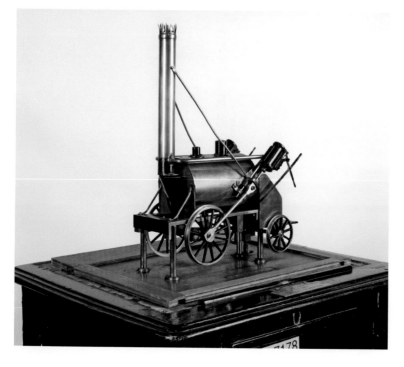

LEFT A penny-in-the-slot *Rocket* model that was used to entertain and raise money on station platforms. This example is in the National Collection at the National Railway Museum.

The railway companies knew that a replica also makes a perfect foil or backdrop to marketing exercises designed to launch new products intended to capture public hearts and minds by association with iconic and much loved images. Consequently the London and North Western Railway, which included the route of the Liverpool and Manchester Railway, was content to use the replica they had built to publicise progress.

Tantalisingly, in the presentation book that Robert Stephenson & Co prepared for Henry Ford in 1929, which is held in the National Railway Museum archives, we are treated to the possibility of six different firebox-shape options that they must have considered, as shown in the memorandum below:

'Memorandum on the Replica of the *Rocket* constructed by
Robert Stephenson & Co, Darlington, Eng. For Henry Ford Esq 1929.'

– so the accurate shape had not been truly established even as late as 1929. It is likely that once the shape had been determined for the first Henry Ford replica, the Crewe replica was simply updated to the flat top shape we see later on in 1937 for the LMS *Duchess* publicity shots shown at the top of page 102.

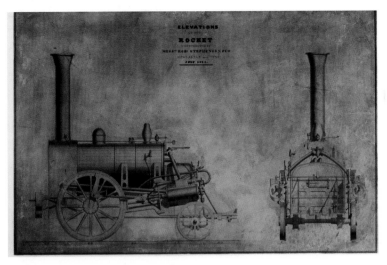

The Henry Ford, R. Stephenson & Co replicas

Henry Ford wanted to mark the centenary of the Rainhill Trials achievement in 1929 by using some of his wealth to create the first of four replicas, built by the Robert Stephenson Company, that could go into museums around the world, to teach the general public not only

ABOVE Side elevation of *Rocket* by Robert Stephenson & Co. Why, if this drawing was available, did their chief draughtsman get the shape so wrong? See page 36.

BELOW Presentation book prepared for Henry Ford by Robert Stephenson & Co, setting out the history of *Rocket* and why the replicas were built as they were.

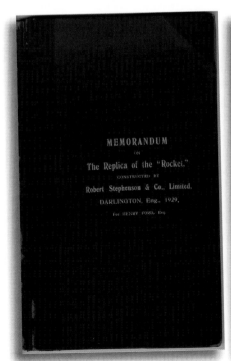

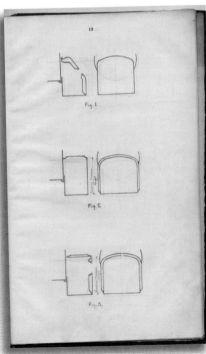

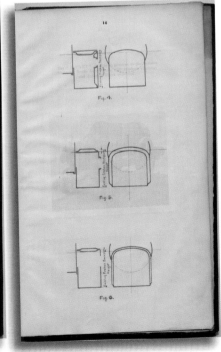

RIGHT London Midland & Scottish Railway succeeded the L&NWR and used the replica *Rocket*, now corrected, to launch their latest express locomotive in 1937.

FOR OVER
ONE HUNDRED YEARS
IN PEACE AND WAR

L M S

IN SERVICE TO THE NATION

BELOW The 1881 replica *Rocket* in rebuilt form finds itself up on stilts outside the new museum at Clapham.

what they looked like but how they worked. Thus two of the four locomotives created on his initiative were sectioned replicas. The fact that the replicas were displayed so often to the public over the intervening years since 1929

shows that they have successfully achieved what Henry Ford set out to do – to tell the story of *Rocket* in a way that links the past with the present. That is the role of a museum, and why the four Rainhill Centenary replicas were explicitly made for museum display. The four engines were Numbers 4, 6, 7 and 8 in the table on page 98.

The 1923 Buster Keaton movie replica

The replica made for the Buster Keaton movie *Our Hospitality*, Number 3 in the table, pre-dates the Robert Stephenson & Co museum replicas, and was a remarkable piece of mechanical set-building that had to function in difficult conditions. The slope of the cylinders was something of a compromise, reflecting the post-L&MR opening version of *Rocket*; but the rest of the model owed its origins to the iconic shape we have come to know and love.

The 1937 rebuilt Crewe replica for the LMS

By the time the London Midland and Scottish Railway (successors to the Crewe dynasty) rolled out their latest Coronation Class top link main line express locomotives in 1937, shown above left, the replica had been rebuilt yet again, with a more correct firebox and rear wheel-set. This made a spectacular foil to set against their most powerful latest locomotive, built for pulling the prestigious 'Coronation Scot' train up and down the West Coast Main Line.

In 1829 John Rastrick estimated the power of *Rocket* at the Rainhill Trials at 22hp. By comparison, the *Duchess* class of locomotive (above left) could sustain 2,200hp provided the fireman could keep shovelling coal into the firebox fast enough! That is 100 times more powerful in 100 years.

In 1961 the updated 1881 replica was placed on a raised plinth outside the new Transport Museum at Clapham, before eventually serving as a donor machine for some of the parts of the 1979 Satow replica.

MUSEUM OF BRITISH TRANSPORT

The 1979 Mike Satow replica

The Mike Satow 1979 replica (Number 9 in the table) was dramatically different from those that had gone before, and those of us who have had the chance to work closely with this replica owe him an enormous debt of gratitude. Here was an instantly visually recognisable working machine that could, in addition, perform like the real *Rocket* did at Rainhill yet within the constraints of modern safety rules. More importantly it gave those familiar with handling 20th-century railway locomotives the chance to get inside the minds of the replica creators as they wrestled with the challenges of making the comparatively crude replica *Rocket* perform in public with all the idiosyncrasies of no brakes and unconventional reversing on a modern railway network, so eloquently described by David Mosley in chapter three.

The 2010 Jim Rees replica

The Mike Satow replica had lasted 30 years in almost continuous service and had travelled all over the world, clocking up many more miles than the original ever did. The need to renew this replica when it became worn out gave the author the opportunity to examine in detail some of the arguments that were put forward by the National Railway Museum in 2008, to override the idea of a straight replacement for the 1979 Mike Satow machine and instead to create a replica more akin to the *Rocket* of the Rainhill Trials, supported by heritage railway benefactor Alan Moore.

The trials and tribulations of designing and building the 2010 replica (Number 10 in the table) are, fortunately, still fresh in the minds of the principal protagonists, so it has been possible to describe them in the next chapter of this publication. In essence the idea of creating a more accurate replica *Rocket* within the constraints of the materials, processes and modern techniques, dictated by the legal requirements of the present day, has created a fascinating dialogue. There were daunting challenges in creating a *true* Rainhill *Rocket* replica, but the 2010 replica is a lot closer to the Rainhill *Rocket*, and creating it has taught us a lot more about how the Rainhill Trials machine functioned in its heyday.

The function of replicas and what we learn from them

We are so fortunate that there have been enthusiastic people like Henry Ford, Mike Satow and, more recently, Alan Moore who have exercised their philanthropic right to fund these recreations of *Rocket* during the last 100 years. The fruits of their investment are that the iconic shape and form of the Rainhill-condition Stephensons' *Rocket* continues to be known, loved and operated around the world.

Thank goodness this is so, because the original *Rocket* in its present state, isolated in the Science Museum as a static exhibit, is so far removed from the 1829 Rainhill crowd-pleaser that this book would have otherwise been a dull story of a long-forgotten machine whose much-altered remnants had simply survived and been researched. It lacked the glamour of the iconic version of Stephenson's *Rocket*. By having the Bailey and Glithero archaeological investigation of the 1829 machine to supplement the Henry Ford, Robert Stephenson & Co replicas and the 20th-century working replicas, our world is that much richer. We can judge the significance of *Rocket*'s place in history and be grateful for the additional knowledge that replicas bring to our understanding of the innovative engineering of the original locomotive.

By the time that Robert Stephenson & Co were commissioned by Henry Ford in 1929 and subsequent years to build four replicas of *Rocket* over six years, the uncertainty over the firebox was thought to have been settled. The first three were destined for museums in the USA, but the whereabouts of the third (Number 6 in the table) is not known. We owe a huge debt of gratitude to Henry Ford, the man who championed mass production in the automobile industry, for his philanthropy and fastidious approach in commissioning these educational masterpieces! Ford was particularly interested

RIGHT One of the
replica *Rockets* being
erected at the works
of Robert Stephenson
& Co in Darlington.
(Northern Echo)

RIGHT AND FAR
RIGHT Furnace-
brazed boiler tube
rolled up from flat
sheet and joined.
(Author)

in the ways that significant artefacts like *Rocket* had been manufactured, and persuaded Robert Stephenson & Co to use old-fashioned techniques to manufacture the centenary replicas. For example, the 25 copper tubes in the boiler would have been available only as solid-drawn copper tube in 1929 – that is, there would have been no brazed seams. So the Stephenson Company were put to the trouble of rolling up copper tubes from flat sheet and brazing them into tubes in order to manufacture the fire-tubes authentically, the way that it had been done in 1829.

This, to me, is what makes replica locomotives so useful to historians of engineering. We might never have found out such things from evidence based on the remains of the original *Rocket*, with all its non-ferrous material long-since vanished. In addition it was agreed by the experts, consulted in the 1920s by the Robert Stephenson Company, that the enigmatic shape of the inside of the copper firebox collar must have been incorrectly drawn by John Rastrick at the Rainhill Trials. The experts said that it must have been bulged on the inside of the firebox as well as the outside. So that is exactly how the four replica fireboxes were made. But Bailey and Glithero proved that John Rastrick was right, and the experts in the 1920s were wrong! The boiler water levels determined in their careful examination clearly show that the inside of the firebox was level and flat, and that Rastrick's flat interior firebox sketch was correct. The story goes on unravelling, in spite of *Rocket*'s replicas conveying a misleading story to people all over the world!

The creators of the firebox for the 1979 replica must have been relieved that they had to make a two-dimensional outer firebox shape and not the strange three-dimensional shape of the 1881 Crewe-built replica *Rocket*. But little did they know that there was another trap awaiting them.

When the Museum of British Transport was set up in London in 1961 to exhibit its collections, the much-rebuilt 1881 replica was displayed on an elevated pedestal at the entrance. By 1975 the dream to create a National Railway Museum at York had become a reality. Happily this coincided with completion of the building by Mike Satow of the replica 1825 *Locomotion* built by Locomotion Enterprises to celebrate the 150th anniversary of the Stockton and Darlington Railway in 1975. His replica took part in the events celebrating the anniversary and led the cavalcade of locomotives on the tracks of the old Stockton and Darlington Railway, which represented locomotive development over the 150 years since George Stephenson's *Locomotion*. The replica proved to be a huge crowd-pleaser, and over 40 years later can still be seen drawing crowds in the Beamish Open Air Museum in County Durham.

Mike Satow was a former engineer and director with ICI in India. He was a passionate historian of all things railway. After retiring to the UK he was the natural choice of project leader to recreate a working *Rocket* replica, having built the successful *Locomotion* replica in 1975. In 1978 he set up a team of young school-leavers (with the help of the Manpower Services Commission) at Springwell on the Bowes Railway in Tyne and Wear. It was an appropriate venue in the North-East for recreating *Rocket*, as the Bowes Railway was originally built by George Stephenson. The team worked under the guidance of Mike Satow and other skilled engineers, and in 1979 the ninth known *Rocket* replica was completed and christened, not with champagne but with a bottle of Newcastle Brown Ale!

LEFT Replica chimney top for the 1979 replica being completed at the Springwell workshops by Manpower Services Commission trainees.

LEFT Firebox of 1979 replica arrives at Springwell for Mike Satow to examine.

BELOW The 93-tubed boiler barrel of the 1979 *Rocket* replica being readied for use.

The boiler created to provide steam for the replica designed by Mike Satow and Whessoe Ltd left those involved with two dilemmas. Firstly, the multi-tubular boiler design originally proposed by Henry Booth was developed by Satow to its logical conclusion. Although the 25 tubes of the Rainhill Trials locomotive gave *Rocket* a huge advantage, Mike Satow knew that it was possible to get even *more* heat transferred into the water by using many more small tubes. He elected to use 96 tubes of 1.25in diameter. This gave his working replica a theoretical advantage over the real *Rocket* of a more than 50% greater tube heating area. Compare this boiler to the images in chapter two of *Rocket*'s 25 tubes versus *Locomotion*'s single tube.

Secondly, in the absence of the evidence from the *Rocket* archaeological investigation some 30 years later, Mike Satow followed what we now know to be the mistaken design of the 1929 Robert Stephenson & Co replicas by bulging the firebox plates both inside and out. The middle picture on page 105 shows Mike Satow pondering the newly delivered steel firebox with its raised inner portion and no doubt wondering how it was going to perform in steam.

So was Mike Satow trying to improve the performance of his replica over the original by increasing the area of the inside of the tubes? We shall never know – but what we do know from 30 years of experience of the 96-tube Mike Satow replica and five years of experience with the more authentic replacement Flour Mill

25-tube replica, is that on damp mornings, when the breeze was from the rear of the locomotive, it was sometimes impossible to get a fire going in the 1979 replica *Rocket* due to lack of draught. The experience of the 2010 replica is that it will light and burn freely and rapidly with no restrictions from damp and direction of wind. Although Mike Satow perhaps believed he could advance the multi-tube advantages argument a bit further in his replica, it was the Rainhill-type 2010 replica *Rocket* that the crews preferred to operate.

There is something else that those of us familiar with these two replicas have noticed. Look at the base of the white chimney in the photographs of the 1979 replica. It is very unusual to see burn marks on the paint, though admittedly there was evidence of scorching on the chimney after the prolonged period of hard steaming that the locomotive endured in the re-enactment for *Timewatch* (which will be described in chapter six).

But the chimney base of the 2010 replica seems to be permanently scorched. This suggests to me that the flue gases emerging from the 1979 replica had given up most of their heat by the time they reached the chimney base. That means the greater number of small tubes were extracting nearly all the available heat from the gases. This is not the case with the later 25-tube configuration, but free passage of hot gases is the bonus earned by wasting some of the fuel. To decide on which replica was better, I would vote for the 2010 configuration. Jim Rees, who project-managed the 2010 replica, needed no convincing. He wanted the replica to be as near to Rainhill condition as possible.

The other dilemma concerns the recurring confusion over inner firebox shape. The Ford/Robert Stephenson & Co replicas all assume that the firebox inner shape replicates the outer shape, and that John Rastrick was wrong in his Rainhill notes and sketches. Mike Satow replicated that very difficult double-curved shape for the inner firebox as defined in the 1930 sectioned version. However, this is where we have travelled full circle in terms of once lost and rediscovered knowledge. The Bailey and Glithero 'dig' in 1999 proves beyond any shadow of doubt that the 1829 firebox was flat

BELOW Replica *Rocket*'s chimney scorched after hard running in the rerun of the Rainhill Trials.

LEFT Replica *Sans Pareil.*

on the inside and contained no inside bulging plate as shown in the new replica steel firebox image. So Rastrick was right after all, and here was something else that the 2010 replica version of *Rocket* could put *right* when replacing the 1979 Mike Satow version.

In addition to creating the *Rocket* replica in 1979 Mike Satow was instrumental in persuading others around him to create two more replicas. He set out to create Timothy Hackworth's *Sans Pareil* and Braithwaite & Ericsson's *Novelty.* His vision was that the three former rivals could once again slog it out in a replay competition. Neither of the two replica rivals to *Rocket* initially acquitted themselves brilliantly in the pageant and cavalcade of *Rocket 150* in 1980, which was held to celebrate the 150th anniversary of the Rainhill Trials. However, they both got a second chance to show their mettle later on when the BBC *Timewatch* programme staged a re-enactment of the Rainhill Trials in 2002. This exciting opportunity will be described in the next chapter.

LEFT Replica *Novelty* arrives from Sweden.

CHAPTER SIX
The value of running with replicas

Replica *Rocket* in steam. *(Don Bowerman)*

Replicas that operate

The 1979 *Rocket* replica along with the *Locomotion* and *Sans Pareil* replicas toured Britain and the world, operating, giving rides and delighting and educating people. The Mike Satow *Rocket* replica has travelled many thousands of miles more than the original locomotive, admittedly mostly in the hold of a Jumbo Jet or on the back of a low-loader! Since the creation of these successful working replica locomotives other initiatives have sprung up, and we now have several machines that fill the missing gaps in railway history, like Trevithick's 1804 *Pen y Darren* replica locomotive.

Jim Rees, keeper of industry at Beamish open air museum, and his colleague Andy Guy did much research to extend our knowledge of early locomotives. This enabled them to produce two further replica early machines to join the *Locomotion* replica on their early railway at Beamish, the Pockerley Waggonway, constructed with historical accuracy so that the public could see replica trains of the early 19th century working within an historical context. Replicas of *Steam Elephant* and *Puffing Billy* were created to join *Locomotion*, and an attractive period locomotive shed was designed and built to house and show off this wonderful collection of early locomotive replicas. This takes the whole concept of working with operating replicas to a new level of public participation, so that an understanding of the storage, maintenance and repair processes can be researched and shared with the public.

The Beamish collection

The three replica locomotives at Beamish museum fill in many of the gaps in our understanding of early locomotive technology before *Rocket* came along and changed everything. The public can enjoy studying the mechanisms of these locomotives in context and ride behind them on the museum's Pockerley Waggonway.

Steam Elephant was a six-wheel, wooden-framed Stephenson gauge locomotive with two vertical cylinders mounted internally within the centre-flue boiler. The original was built by Chapman and Buddle in 1815. As with the later *Locomotion,* piston rods emerging from the top of the cylinders drove connecting rods that in the case of *Steam Elephant* turn two crankshafts between the three driven axles through reduction gears between the frames. It weighed about seven tons and could pull 90 tons at 4mph on the Wallsend Waggonway.

Puffing Billy was made for Wylam Colliery by William Hedley in 1814. Orginally, it had four wheels and was later converted to eight wheels to spread its weight better on early rails. It was conveted back to four wheels after the Wylam track had been changed to edge rails. It survived in working order

BELOW Replica locomotives at Beamish Museum. Left to right: *Puffing Billy*, *Locomotion* and *Steam Elephant*. *(Beamish Museum)*

until 1862. The original is on display at the Science Museum in London along with *Rocket* in its 'Making the Modern World' exhibition. The replica was built in 2006.

Locomotion was started in 1824 to George Stephenson's design for the Stockton and Darlington Railway. It had a single centre-flue boiler with two vertical cylinders half-buried in the boiler, driving on to the four wheels through yokes and connecting rods. It was the first locomotive completed in 1825 at the new Robert Stephenson & Co factory in Newcastle. It was also the first locomotive built for a passenger-hauling railway. The replica was built by Mike Satow in 1975 to celebrate *Locomotion*'s 150-year-old design. The original locomotive is on loan to Darlington Railway Centre and Museum and is displayed as part of the National Collection.

The 2002 BBC *Timewatch* recreation of the Rainhill Trials

In 2002 the opportunity arose to do something very special with operable early locomotive replicas. The BBC approached the National Railway Museum to see if it would be possible to film a re-enactment of the Rainhill Trials with the replicas of the three original principal contestants. Although *Rocket* and *Sans Pareil* replicas were available to take part, the *Novelty* replica had by this time been passed on to the Swedish Railway Museum as a static exhibit, because it had never operated satisfactorily.

The BBC *Timewatch* producer, Deborah Perkin, proposed that the BBC fund the overhaul of *Novelty* in the well-equipped engineering workshops at the National Railway Museum, and modify *Novelty* as necessary to turn it into a viable competitor, provided the director of the Swedish Railway Museum, Robert Sjoo, gave his permission. He enthusiastically jumped at the possibility of rewriting the history books in favour of a Swedish victory! Consequently he gave the National Railway Museum a letter of authority that gave our engineering team at York a free hand to do whatever we thought was

necessary. As chief engineer of the project I felt honoured that Robert Sjoo had faith in our ability to do the best possible job.

First it was necessary to establish what was wrong with the 1979 *Novelty* replica. Why had it always fallen off the tracks when attempts had been made to operate it? The services of the National Railway Museum's friend and early locomotive expert John Glithero were sought. He agreed to go to the Railway Museum at Angelholm in Sweden where the replica was on display and to investigate the likely problems in resurrecting the machine as a viable running locomotive.

John's report provided many detailed observations. Amongst them were that the spoked wheels on which the *Novelty* replica stood were both too narrow for modern track-work

TOP Replica *Steam Elephant* at Beamish, with its creator, Jim Rees, on the footplate. *(Beamish Museum)*

ABOVE Replica *Puffing Billy* outside the replica engine house on the Pockerley Waggonway at Beamish Museum. *(Beamish Museum)*

LEFT Replica *Rocket* at speed during the re-enacted Rainhill Trials.

CENTRE LEFT Model of *Novelty* in the Science Museum that may have been used to determine the wheel thicknesses for the replica.

and too wobbly. It turned out that many of the dimensions for building the full-size replica were obtained from scaling up a beautifully made model that existed in the collections of the Science Museum. The builder of that model, possibly the model shown left centre of this page, had been forced to make assumptions about dimensions where no accurate information was available. The result was that although the distance between the wheels was correct, the overall width of each wheel was effectively some 1.75in (40mm) too narrow. Although this might seem a trivial amount, the geometry of modern track-work is such that if wheels are too narrow they will drop into the flange clearance gap at points and crossings. This discovery added to the finding that straight-spoked bicycle-type wheels were insufficiently strongly braced to withstand side loads, which meant that we had our explanation of why the *Novelty* replica would not stay on the track. Although the wheels looked like spoked bicycle wheels, the spokes did not cross over each other sufficiently to provide rigidity in the manner of properly spoked bicycle wheels. Other than that John's report was optimistic about the chances of repair. We all knew the boiler had done very little steaming in the intervening 20 years.

Preparing the *Novelty* replica to run

The engineering team at the National Railway Museum was confident that these problems could be addressed. The plan was to extend the treads of the wheels with rings of the right diameter attached to the face of the tyre, as well as to tighten all the spokes within the hubs to make the wheels resist side loads more effectively. The boiler, which had not been

LEFT Replica *Novelty* arriving at the National Railway Museum workshop from Sweden.

steamed for many years but was relatively unused, would be recommissioned. Reassurance was given to the BBC's *Timewatch* makers that the museum could sort out the problem in the time-scale proposed and a budget figure was agreed. The BBC arranged for the locomotive to be transported from Angelholm and all the necessary port handling and temporary export documents to York were put in place.

My initial proposal (and costing) was based on making the new wheel extensions out of lengths of 40mm square bar being rolled into circles of the right diameter and welded at the join to make a perfect circle. However, this was to prove to be an over-optimistic solution. Bending steel (or even plasticine) into a circle thickens up the inner edge and causes the faces to be no longer square to the axis. None of the possible sub-contractors that we approached were willing to make the tyre extensions this way. They wanted to machine the new tyres from solid pieces of steel plate, taking the proposed costings that we had agreed with the BBC way over budget. The BBC, understandably, did not want to extend their budget. I was fortunate that my relationship with the Friends of the National Railway Museum was sound enough for me to be able to explain my predicament to them and ask if they could help me fund the more expensive wheel extensions in exchange for a credit on the BBC programme. Both parties were agreeable to this solution and an engineering firm in York speedily produced four satisfactory rings.

Fastening the rings to the existing wheels was done with high tensile bolts using a magnetic base drill to drill through the new ring and old wheel in 16 places. Then the new ring was opened out to the bolt diameter and the old wheel rim was threaded to take 16 concealed cap head bolts of 12mm diameter. These were counter-bored through the new tyre extensions and threaded into the old wheel rims. Once the bolts were all in tightly, each counter-bored hole in the new rims was filled with an oak plug hammered in and faced off. By the time that the paint was applied to the wheels, the repair was invisible. All the spoke nuts were tightened by an equal amount, one flat at a time, and the wheels checked for running true.

That just left the problem of the boiler. Our boiler insurance company needed to certificate

ABOVE New wheel rims arriving at York ready to be fitted to *Novelty*'s narrow wheels.

the boiler as safe to operate. The original certificate had long expired. The boiler with its peculiar shaped, enclosed firebox and unique horizontal return flues was dismantled and examined. The boiler inspector highlighted several issues that needed attention before he would consider giving it a certificate. Firstly there were two tapped holes in the boiler shell for try cocks, as previously described for *Rocket*. These were threaded straight into the thin boiler shell. They needed steel bosses welding on to the boiler barrel to make them robust and fit for purpose.

Secondly, the boiler inspector found that one of the large tapped holes round the foundation

BELOW Tyre-fastening the new rims on to the old wheels using concealed cap head bolts flushed off with oak plugs.

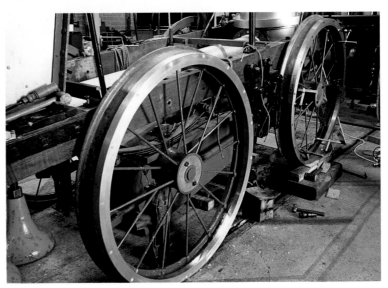

RIGHT Condemned tapped holes in *Novelty*'s boiler needing bosses to be attached.

FAR RIGHT Steel bosses ready to be attached to boiler by coded welder to gain approval from the boiler inspector for replica *Novelty*'s boiler.

RIGHT Condemned tapped hole in replica *Novelty*'s foundation ring at base of boiler.

RIGHT Ground-out defect in boiler barrel of replica *Novelty* ready for welding.

RIGHT Defect in replica *Novelty*'s boiler welded-up by coded welder.

RIGHT Screwed plugs in the three pass return flues in the horizontal section of *Novelty*'s boiler.

ring was incorrectly tapped out and the thread was damaged and potentially unsafe. This needed an insert fitting and rethreading.

Lastly, he used magnetic particle and dye penetrant non-destructive testing to discover a joint seam that had not been properly welded during construction. This was ground out and re-welded by the use of a certificated welder. Amazingly, we also found that the fusible plug – a legally required lead-filled safety device which melts when the firebox goes over temperature and extinguishes the fire should the water level drop dangerously low – had been inserted originally from the wrong side, and gave no protection to the boiler whatsoever.

In addition we found that it was almost impossible to clean out the three-pass return bend steel pipe flues that passed through the horizontal water spaces, providing the exhaust passage from the fire to the chimney of the locomotive. We were able to fit screwed plugs in the ends of the flues instead of return bends so that at least a vacuum cleaner could be inserted into the open-ended pipes. This turned out to be a maintenance nightmare for keeping the exhaust passages free of debris. At one point we wondered whether it might be possible to use a firework dropped down the chimney as a possible way of keeping the tubes clear!

Eventually, when all the inspections had been done and the boiler inspector had pronounced himself happy with the work, the boiler was reassembled. A satisfactory hydraulic test to one-and-a-half times the working pressure was carried out with no leaks or problems. We were greatly relieved.

All this reconstruction and repair work was carried out in full view of the public visiting the museum. We had learned from the *Rocket* 'dig' that the public are interested in engineering history and skills. Consequently museum visitors to the workshop balcony were treated to regular blackboard updates explaining progress, and regular 'Rocket Round-ups' explaining the historical and engineering implications of what we were doing.

As the work to reassemble *Novelty* was taking place I developed a strong bond with *Novelty* and its 1829 crew, John Braithwaite and John Ericsson. Members of the public started asking us which was the better of the two machines and would *Rocket* still win the soon to be restaged Rainhill Trials? At that stage it had never entered my mind that the outcome of the trials could be anything other than an outright win for *Rocket*, no matter how good a job our engineering team had done on *Novelty*.

With *Novelty*'s slide valves set to the best compromise possible between forward and reverse settings, it was time to try out the locomotive at the rear of the National Railway Museum away from the public's gaze. From the minute of lighting the fire it was obviously a lively machine that liked a struggle!

The firing technique was to drop fuel down the chimney at the top of the boiler into the bed of red-hot fire, doing your best to aim it so that the circular grate got a chance of an even covering. Steam was soon raised, and after a few safety checks and the tightening up of fittings and flanges where a few initial leaks were found, it was time to take *Novelty* for a spin and see if it would stay on the tracks.

We need not have worried. The new extended wheels behaved perfectly and the locomotive negotiated the gaps in the modern points and crossings with ease. However, there were two immediate snags apparent as *Novelty* moved off. Firstly, the exhaust steam was directed down to the floor from the cylinders with a loud explosion for each cylinder stroke. This noise was such a contrast with *Rocket*'s gentle *woof* as it exhausted the spent steam and induced the draught to the fire. Secondly, the set of bellows, which was designed to offer a forced blast to the underside of the fire grate, was only operating for one direction of its stroke. In other words it was not a double-acting bellows but single-acting. Had the bellows operated from both sides of the wind box things might have turned out differently, with the forced draught more of a constant blow

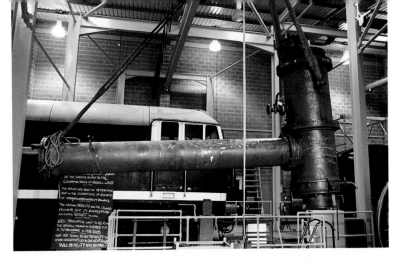

rather than being applied in discrete pulses. To this day we still do not know whether the real *Novelty* had leather bellows on both sides of the centre of the wind box. My guess is that it did. I sometimes wonder why Braithwaite and Ericsson did not think of directing the exhaust blasts upwards over the end of the chimney to improve the draught of the fire and soften the staccato blasts from the cylinders.

The team that had volunteered to drive and fire the locomotive in the re-enactment were invited to come and get some practice in on the hidden sidings at the National Railway Museum. They were Richard Lamb (an existing National Railway Museum volunteer) and his friend Andrew Hurrell. They were keen to get to know their charge with all its many foibles.

We at the National Railway Museum were satisfied that we had done our best to create a viable machine that could take part in the contest. I must admit that there were a couple of times during the rebuild when I felt that by doing the best possible job we might be sowing the seeds of *Rocket*'s defeat in the re-enactment! But in my mind I knew that *Novelty*'s constrained horizontal flue gas exit and the steam exhaust discharging straight down to the ground from the cylinders were never going to be winning features.

115

The re-enactment of the Rainhill Trials

*N*ovelty's limited steaming trials were completed. *Rocket* was given a final check over and adjustment, and both locomotives were packed off on a low-loader to join the replica *Sans Pareil* for a week of filming and fun on the Llangollen Railway line, at a delightful place in the Dee Valley called Carrog. This site was chosen because it was one of the few places on a preserved heritage line in the UK where it was possible to replicate the track conditions found at Rainhill: that is, level track for the same length, with the ability to define acceleration and deceleration sections at each end of the measured mile. The track also had good road access at the Carrog site for road transport of the vehicles.

Three early railway expert judges were appointed to oversee fair play and to do the calculations for the re-enactment. Who better than the former *Rocket* 'dig' team, Dr Michael Bailey with Dr John Glithero, officiating as judges with their work colleague Peter Davidson? The team

TOP Team replica *Rocket* ready for the Rainhill Trials re-enactment. *(Martyn Stevens BBC)*

LEFT Team replica *Sans Pareil* ready for the Rainhill Trials re-enactment. *(Martyn Stevens BBC)*

LEFT Team replica *Novelty* ready for the Rainhill Trials re-enactment. *(Martyn Stevens BBC)*

BELOW Judges limbering up for their arduous duties. Left to right, Dr John Glithero, Dr Michael Bailey and Peter Davidson.

of judges installed themselves in a brake van at Carrog station overlooking the start of the route.

The replica trials followed the same procedures as the original trials in 1829. There was an opening ceremony at which the chief judge Dr Michael Bailey made the rules clear to the competitors and the audience in the grandstand. To everyone's surprise, pleasure and amusement, the *Timewatch* team had secretly constructed a full-size replica of Brandreth's *Cycloped* complete with replica horse, which was pushed into view for the opening ceremony as a possible fourth competitor. There was little chance of the plastic horse achieving the ten miles per hour average speed called for by the judges! The replica was donated to the National Railway Museum by the BBC as a thank-you to the museum for getting the *Novelty* replica operational again.

The three locomotives gave demonstration runs on the first day. Individual performances for

RIGHT Replica *Novelty* on test.
(Martyn Stevens BBC)

ABOVE Replica *Sans Pareil* on test.
(Martyn Stevens BBC)

all three locomotives were carefully timed on each of the following three days, with their calculated weights being carried in the carriages attached to the locomotive. Special attention had been paid in the design and construction of the replicas to overcoming the known weaknesses of the original machines in 1829, which meant that the defects that produced such disastrous performances on that occasion would not be repeated.

Sans Pareil went first, and put up a spectacular and surprisingly fast performance, even though it used a great deal of fuel. It completed the trials competently without breaking down. The next day it was *Novelty*'s turn and the locomotive performed splendidly, demonstrating both speed and economy of operation. However, *Novelty* suffered from excessive clinker formed on the fire-bars. As the grate gradually became more choked, the clinker prevented the fire performing consistently throughout the day. Unfortunately for the team driving *Novelty* the grate could not be cleaned in situ without dropping out the whole fire and relighting the furnace. This elaborate process was not an option at the halfway break and so the *Novelty* replica had to withdraw from the contest.

The pressure was on for *Rocket* to perform well on the third day and after good consistent running in the morning the crew found themselves fighting against loss of steam pressure as the mid-point of the contest was reached. Again, it was clinker forming on the fire-bars and choking the capacity of the fire to produce heat that was causing the loss of power. Fortunately for the *Rocket* team the grate is fully accessible from inside the ash pan by a crew-member lying underneath the locomotive with a long poker and pushing upwards below the fire-bars. After a frantic half-hour of grate cleaning the clinker fused on to the fire-bars at the morning break was removed and the locomotive was ready for action again. *Rocket* went out and performed even better in the afternoon than it had in the morning, as the crew's confidence grew to achieve what seemed at the time to be alarming speeds.

The judges retired to their brake van to deliberate. They had to adjust their calculations for the differences between the *Sans Pareil* replica and the original. It was the same with *Rocket* with its more modern boiler (96 tubes instead of 25 in the original at Rainhill).

The judges' final figures were surprising. The efficiency of *Novelty* was equivalent to that of *Rocket*, and *Sans Pareil*'s spritely performance was not what the pundits were expecting. Although a lot of fuel was used on *Sans Pareil*,

BELOW Replica *Sans Pareil* on loaded test run.
(Martyn Stevens BBC)

the crew were able to keep their fire in good condition throughout the day and keep their steam production consistent. Those of us representing *Rocket* were aware that there were some amongst the crews who would have liked the outcome of the 2002 *Timewatch* trials to be different from the 1829 Trials. Certainly it might have made better television. But there was in the end no doubt about the best locomotive in both sets of trials as *Rocket* was declared the winner. The National Railway Museum team was able to breathe a collective sigh of relief, whilst congratulating the outstanding performances from their rivals, whose competence outshone the Rainhill Trial results with their original locomotives. It was great to win but for me the real achievement was in the value of being able to operate competent replicas to inform, entertain and educate a wider audience through the medium of television. Everyone participating had a great time and went away satisfied and pleased.

The team from the Swedish Railway Museum who owned the *Novelty* replica were delighted to have an operating replica at last. They were already looking forward to taking the machine back to Sweden to show it off to John Ericsson's fellow countrymen. They also hatched a plot to bring the *Rocket* replica over to Gävle in Sweden to join the operating *Novelty* and many other early locomotives for a festival in 2009. Sweden is significant throughout the international railway museum world for its spectacular collection of early locomotives from the mid-19th century, many of them British exports, and Gävle was the perfect place to have a celebration of many significant operating locomotives.

By the time of the festival the Swedish crews of the replica *Novelty* had become very proficient at managing their locomotive, so *Rocket* and *Novelty* were able to compete against each other for the entertainment of crowds once again. A postscript received from Robert Sjoo, the Director of the Swedish Railway Museum at Gävle, acknowledged the achievement of returning the replica of *Novelty* to running condition: 'Thanks to the BBC and the National Railway

ABOVE *Rocket* and *Novelty* replicas resting in the company of early Swedish locomotives, many built in Britain. *(Heinz Voregger)*

BELOW Replica *Rocket* faces the adoring crowds at Gävle in Sweden.

119

Museum we had the pleasure to be involved in a very exciting adventure. Actually we did not know what to do with the *Novelty* replica we had received in the early 1980s from a local foundation. It did not work well enough for running and as an exhibit I considered it a bit of a too hasty work. It was even on loan to the recently opened Museum of Rail and Track in southern Sweden. When the BBC contacted me for the *Timewatch* programme the value of John Ericsson's pioneer work became clear once again. (Here in Sweden, Braithwaite is very seldom mentioned.) The National Railway Museum team did a superb job to make the replica work and to make it safe for the revenge meeting. I had no problem with the changes to the wheels that were necessary for the permission to run.

'Of course *Novelty* lost again! Neither the original nor the replica was well enough built to challenge *Rocket*, but it was fantastic to take part in the popularisation of the essential railway history for British TV viewers.

'Unfortunately Swedish Television didn't show any interest, which is a common problem concerning the heritage from the railways in this country. However *Novelty* had become a star in my eyes when she returned home. We did events at our museum as well as in Stockholm. The highlight was when *Rocket* Replica and the National Railway Museum team led by Richard Gibbon visited us in Gävle for a great steam event in 2009. For the pleasure of the Swedish audience, we agreed upon a draw that time, and *Novelty* since then is a part of the permanent exhibition of the Swedish Railway Museum. The replica has its own story now!'

The 2010 replacement *Rocket* replica

The 1979 *Rocket* replica was not allowed to rest on its laurels after the *Timewatch* re-enactment of the Rainhill Trials and trip to Sweden. The boiler was by then over 30 years old and was difficult to inspect for corrosion in hidden parts. Everyone agreed that a new boiler was needed if the replica was to continue operating. The National Railway Museum's Curator of Rail Vehicles, the same Jim Rees who had created the Beamish early replica locomotives, started to toy with the idea of the

new boiler being a much closer copy of the 1829 version, with the correct number of tubes of the right diameter. In 1979, for his own replica design, Mike Satow had used a boiler shell which was modern in its construction and almost off-the-peg for the manufacturers, the Darlington-based heavy engineering firm Whessoe Ltd. Jim Rees saw the opportunity to push early locomotive knowledge further by replicating the original *Rocket*'s 25 three-inch diameter tubes and creating a new copper firebox that drew on the findings of Bailey and Glithero's 1999 'dig'. It was important to the designers that the firebox should be made of copper as with the 1829 version to improve the heat transfer from fire to water, as the Stephensons had intended. Graham Morris, a noted steam locomotive engineer, was commissioned to design the new boiler.

Also, if the 2010 replacement replica boiler was to be built to reflect the Stephensons' original design it could not just be a new boiler. The locomotive's Z-shaped supporting frames needed to be radically altered or renewed to accommodate the depth of the accurate firebox and barrel. The trailing wheel set that had graced the Satow 1979 *Rocket* replica had the incorrect number of spokes and was the wrong diameter. Further replacement components were necessary to create a much closer replica version of *Rocket* to the original. The project was becoming a goalpost-moving nightmare!

Here in Jim Rees' words, when he came to set the specification for the new replica in 2007, are his reflections on the challenges ahead:

'The overhaul of the replica "Rocket"
'It is a curious moment when a replica locomotive, that one somehow thinks of as forever "new", comes to need a major overhaul, but that is exactly where the "Rocket" replica has got to, and just like the museum's older and more complex locomotives, like "Flying Scotsman", the process is throwing up new debates and new discoveries.

'The working replica of Robert Stephenson's original Rainhill winner of 1829 has now been working for over twenty five years, perhaps begging the question, "when does a replica locomotive become a real one?" It has fascinated museum visitors and the general public all over the country, as well as in America, Australia, Japan, Switzerland and Holland. Running for so

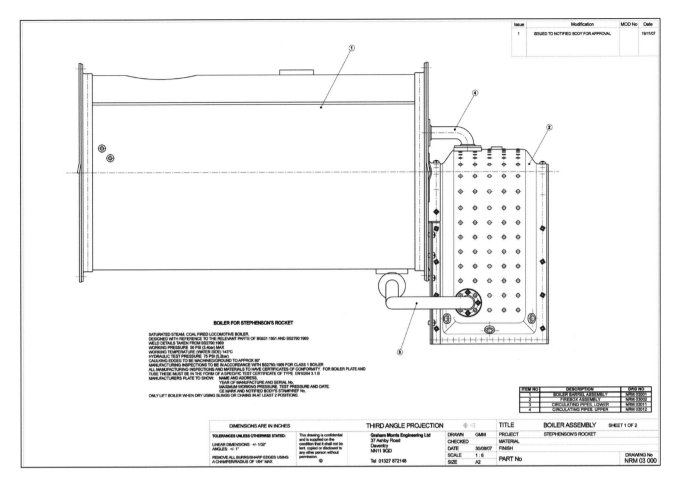

Issue	Modification	MOD No	Date
1	ISSUED TO NOTIFIED BODY FOR APPROVAL		19/11/07

BOILER FOR STEPHENSON'S ROCKET

SATURATED STEAM. COAL FIRED LOCOMOTIVE BOILER.
DESIGNED WITH REFERENCE TO THE RELEVANT PARTS OF BS931:1951 AND BS2790:1969
WELD DETAILS TAKEN FROM BS2790:1969
WORKING PRESSURE 50 PSI (3.4bar) MAX
WORKING TEMPERATURE (WATER SIDE) 143°C
HYDRAULIC TEST PRESSURE 75 PSI (5.2bar)
CAULKING EDGES TO BE MACHINED/GROUND TO APPROX 80°
MANUFACTURING INSPECTIONS TO BE IN ACCORDANCE WITH BS2790:1969 FOR CLASS 1 BOILER
ALL MANUFACTURING INSPECTIONS AND MATERIALS TO HAVE CERTIFICATES OF CONFORMITY. FOR BOILER PLATE AND
TUBE THESE MUST BE IN THE FORM OF A SPECIFIC TEST CERTIFICATE OF TYPE. EN10204 3.1.B
MANUFACTURERS PLATE TO SHOW: NAME AND ADDRESS.
                                YEAR OF MANUFACTURE AND SERIAL No.
                                MAXIMUM WORKING PRESSURE, TEST PRESSURE AND DATE.
                                CE MARK AND NOTIFIED BODY'S STAMP/REF No.
ONLY LIFT BOILER WHEN DRY USING SLINGS OR CHAINS IN AT LEAST 2 POSITIONS.

ITEM NO	DESCRIPTION	DRG NO
1	BOILER BARREL ASSEMBLY	NRM 03001
2	FIREBOX ASSEMBLY	NRM 03002
3	CIRCULATING PIPES, LOWER	NRM 03011
4	CIRCULATING PIPES, UPPER	NRM 03012

DIMENSIONS ARE IN INCHES		THIRD ANGLE PROJECTION		TITLE	BOILER ASSEMBLY	SHEET 1 OF 2

TOLERANCES UNLESS OTHERWISE STATED:	This drawing is confidential and is supplied on the condition that it shall not be lent, copied or disclosed to any other person without permission.	Graham Morris Engineering Ltd 37 Ashby Road Daventry NN11 9QD Tel 01327 872148	DRAWN	GMM	PROJECT	STEPHENSON'S ROCKET
LINEAR DIMENSIONS: +/- 1/32" ANGLES: +/- 1°			CHECKED		MATERIAL	
			DATE	30/08/07	FINISH	
REMOVE ALL BURRS/SHARP EDGES USING A CHAMFER/RADIUS OF 1/64" MAX	©		SCALE	1:6	PART No	DRAWING No NRM 03 000
			SIZE	A2		

many miles, and so often, it is no surprise that it is due for a major overhaul and rebuild, just as working engines always have through history.

'Following the end of its boiler certificate in March, and with the Museum Workshops fully engaged, the loco will be heading south, to Bill Parker's Flour Mill workshops in the Forest of Dean, for the work to begin.

'Built at the request of Dame Margaret Weston, the then Director of the Science Museum, by Mike Satow's team at Locomotion Enterprises, the loco was built, as modestly as possible, with the firm intention of being, not just a valuable interpretation tool, but a "good runner".

'Like all replicas, that involved compromises, not just those, as always, to meet modern regulations and material availability, but those created by the level of historic knowledge at the time. As you may recall, the original engine in South Kensington, is much altered, and incomplete. The wonderful archaeological "dig" of the locomotive, by Bailey and Glithero, had not yet taken place. We know much more about the original now than we did then.

'Mike's wonderful "can do" approach gave other surprises, little remembered today; a non-working replica, built by Crewe around 1880, and long broken up, provided the frames (almost certainly wrought iron), as well as the trailing wheel set, and probably the tender wheels.

'Unfortunately these were, we now know, the wrong size, but they forced compromise dimensions on to the boiler...

'The boiler itself was built as cheaply as possible, and largely welded. The years of use, our modern knowledge, and modern codes of boiler construction, require a new one. This time, though, new and accurate frames, and to a much more authentic, and riveted, design will be provided. We now understand the historic design of the all-important copper firebox, and this we intend to build, as it was, in riveted copper.'

Jim Rees
Curator, National Railway Museum
Rail Vehicle Collections
25/4/07.
(Source National Railway Museum *Rocket* replica technical file)

ABOVE Graham Morris' design for the new boiler for the replica *Rocket* in 2007.

LEFT **Alan Moore, the benefactor who financed the new realistic boiler for** *Rocket*. *(Jimmy James, Bodmin and Wenford Railway)*

CENTRE **Replica** *Rocket* **starting to take shape at the Flour Mill in the Forest of Dean. The tube pattern is clearly seen.** *(Bill Parker)*

The 2010 replica brought its own share of problems to the team that was creating it. First of all there was the question of funding. The Railway Heritage movement in Britain has been blessed with a number of generous benefactors. Henry Ford, of course, was one such benefactor in the past. He had led the way with the four Robert Stephenson replicas in 1929 and beyond. A modern day benefactor who has done a huge amount to enable exciting new projects in the Railway Heritage sector to take place is Alan Moore. Alan sponsored a great deal of work that had been undertaken at the Flour Mill in Gloucestershire with its manager Bill Parker, including several restorations of significant locomotives from the National Collection. Following the specification drawn up by Jim Rees with the new boiler design by Graham Morris, Alan kindly put up the money for the Flour Mill to create a new replica realistic boiler. The Flour Mill generously provided the new frames and other benefactors stepped in to provide the replacement *Rocket* replica with a proper braking system and other accessories that needed updating to make the replica fit for purpose for modern safety requirements.

The new proposed boiler was so unconventional that it needed special dispensation from the National Railway Museum's statutory insurance company. All copper used in steam boilers in the 21st century has to be specified to be type C107. The code number relates to the arsenic content and chemical composition of the copper, so special copper plate and rivets had to be obtained.

Geoff Phelps and Bill Parker at the Flour Mill had other problems to overcome. Instead of a

LEFT **Final redesign of replica** *Rocket*'s **new frame-strengthening system, to avoid putting strain on the copper sections of the boiler.** *(Bill Parker)*

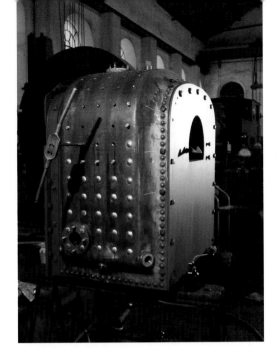

rigid boiler firebox assembly such as that with which the 1979 replica was built, the frame had to actually fully support the soft copper to steel interface at the joint so that there could be no movement between barrel and firebox, and that required considerable frame modification for which there was no precedent. Several designs were suggested and the one shown in the photograph at the foot of page 122 was settled on.

ABOVE LEFT Copper firebox stay-holes being tapped through both thicknesses of copper plate. *(Geoff Phelps)*

ABOVE Geoff Phelps riveting over the heads of the threaded copper stays supporting the firebox walls of the 2010 replica *Rocket*. Another copper stay not yet riveted over can be seen poking through the outer casing, showing the threaded portion. *(Bill Parker)*

BELOW Replica *Rocket* ready at last. Trying it in steam. *(Anthony Coulls)*

Finally *Rocket* Replica 2010 was fit to test, to appear in public and to start its new life as the latest incarnation of Stephenson's *Rocket* in Rainhill condition. It is replica Number 10 in the table on page 98.

Cautionary tales!

The opportunity to delight crowds by running working replicas is something special. However, there are as many problems with operating replicas as with designing and making them. What follows are a couple of accounts of incidents to illustrate the perils awaiting the unwary.

ABOVE Yes! But will it cook a breakfast as well as the old one did, wonders Matt Ellis.
(Anthony Coulls)

RIGHT The 2010 replica *Rocket* makes its debut at the Albert Memorial.
(Andrew Scott)

Rocket and the 'free' repaint

In 1992, a large railway works in the south of England invited *Rocket* and coaches to run at its annual charity open weekend, and give passenger rides to the visiting public on a specially closed-off siding adjacent to the works. During the negotiations with the National Railway Museum for the hire costs the organisers were horrified to discover that not only would they have to pay for transport but also a commercial hire rate for the locomotive and carriages. They managed to get round the question of the transport cost by persuading the local Territorial Army unit that moving *Rocket* from York to Kent and back would be an excellent logistical training exercise for their team. That still left the locomotive and carriages' hire cost to be found. But the organisers of the event had discovered that after this hire, the locomotive was to be temporarily withdrawn for replacement of its wooden cladding and a complete repaint. Ever resourceful, they asked us if their craftsmen at the works could replace the kiln-dried shaped tongued and grooved pine outer lagging (already purchased by the National Railway Museum in readiness), and repaint the locomotive to our usual high standard in lieu of the hire costs. Here was the perfect quid-pro-quo deal that made everybody happy (or so I thought...).

The Territorial Army transported *Rocket*, tender and carriages two weeks before the charity open weekend, allowing sufficient time for the recladding and repainting. Frequent phone calls kept us informed of progress during these two weeks. Glass fibre matting was installed below the steel crinoline bands surrounding the boiler in three places where the wooden lagging slats were supported 1.5in clear of the boiler. The new wood was pre-undercoated and fastened on to the crinoline bands as we had stipulated. It was then given several undercoats and two topcoats of sunflower yellow and two final coats of varnish. When David Mosley (National Railway Museum's education officer) and I arrived on the Friday evening before the event, and saw the locomotive for the first time, it shone like a new pin. There were smiles all round as we pushed *Rocket* and tender out of the workshop ready to fill up with water and light up for a trial run.

The smiles did not last long. All was fine for the first hour after filling and lighting up, but

as soon as the warmth started to heat and compress the air trapped above the water in the boiler, generating a small amount of pressure, we noticed a trickle of hot water from the bottom of the new cladding. We convinced ourselves that perhaps there could have been moisture in the fibreglass insulation that was being driven out. But as the boiler got hotter the trickle became a stream and the stream became a torrent. What was mystifying was that there appeared to be five hot spots in various random places under the lagging. The fire was knocked out and an emergency meeting was convened.

Whatever solutions were suggested, they would have to conform not only to the law but also to the high standards that the National Railway Museum demand and get from their staff and equipment. To withdraw from the operation at this stage was unthinkable after all the effort that had been invested in the event. I identified the hottest of the five patches. This was likely to be the most drastic example of whatever it was we were about to find. I set to with my trusty pocket multi-tool to make a hole in the brand new paintwork, tongue and groove woodwork and fibreglass. By the time we had created a hand-sized hole on the left-hand side we were getting showered with hot water everywhere. As we cleared away the debris we could see that one of the self-tapping screws that held the wooden lagging on to the crinoline was not the regulation 0.5in long; it was fully 2.5in long and had clearly penetrated right into the boiler barrel! The hot water was leaking out down the screw thread. We realised that some of the holes went right through into the water space! We were disbelieving of what we could see at this stage. I remember quietly cursing those who kept reminding me that I was so lucky to have landed the *best job in the world*.

Removing the first screw produced the dramatic waterspout shown in the photographs on page 126, which someone had the presence of mind to capture. The facial expression says it all! Now we realised that the cause of the problem was that the lagging fastening screws had penetrated the boiler in at least five places. Solving the problem presented us with two challenges. Firstly, how could we carry out a safe repair in time to run for the charity event the next morning? Secondly, we were anxious that there might be other screws that were just on the verge of penetrating the boiler. Might they yield when the boiler was up to pressure? Fortunately closer inspection showed that most of the self-tapping screws that had been used to hold the wood on were of the 0.5in type.

We were also fortunate that the National Railway Museum enjoyed a trusting relationship with our insurance company boiler inspector. I rang him at home on the Friday evening apologising for the intrusion and told him of our sorry plight. I proposed that we were intending to drill and tap the five offending holes, before plugging them with taper threaded plugs of preferably one-eighth BSP. He approved the repair on a temporary basis, provided that when the locomotive returned to York the plugs would be removed for his inspection; then the boiler would have to pass a fresh hydraulic test. I

remember thinking as I was speaking to him, 'It's all very well talking like this, but where on earth do I imagine I'm going to find BSP taps and five taper plugs at this time on a Friday evening?' I was reckoning without the kindred spirits of the model engineering fraternity in the area! Inevitably someone in the works knew a model engineer, who we phoned, and he put us on to the local model engineering boiler tester. This was a man after my own heart. He had everything we wanted and more, and was willing to share it with us in exchange for a chance to drive *Rocket* in the morning, if all was well!

Back at the site the five offending screws were drilled out, and we threaded the holes in the boiler barrel with the borrowed taps, taking care to ensure that the taper plugs went in no further than was safe. We used graphite paste on the threads to ensure leak-proof joints. We soon had the hosepipe back into the still warm boiler, re-laying and lighting the fire as soon as the water level appeared in the gauge glass. The pressure rose till we got the safety valves to lift with no sign of any leakage at any of the five repairs. We were able to steam *Rocket* replica once in the pitch darkness up to the end of the demonstration track, and back again, before dropping the fire and stabling the engine prior to adjourning to the nearest pub seconds before closing time for a well-earned reward!

We gave over 2,500 rides to members of the public in beautiful weather over the next two days, and apart from a few scowls from some people who wondered how and why we had spoiled part of that lovely new paint job, hardly anyone had a clue how close we had been to letting everyone down that weekend.

Rocket and *The Day the World Took Off* filming

The second cautionary tale concerns the hiring of replica *Rocket* to take part in a Channel 4 dramatisation called *The Day the World Took Off* in 1998. The storyline involved *Rocket* as the catalyst for the start of the modern world and concentrated on the opening day of the L&MR. The filming took place on the East Lancashire Railway near Bury. David Campey was the National Railway Museum's engineer/locomotive driver.

BELOW Self-tapping screws right into the boiler! The boiler-leak crisis. The stream of hot water can be seen just above the author's elbow, a worrying moment for the author and colleague David Mosley, whose account of driving the *Rocket* replica appears on page 81. *(Author)*

RIGHT *The Day the World Took Off*. Dave Campey and his fireman prepare for their drenching during filming. *(Denis Bradley)*

There were two memorable scenes worthy of recall. In the first, the producer required *Rocket* and its empty carriages to appear from under the bridge (at Bolton Street station in Bury) and pull up at exactly the right spot where the Duke of Wellington, Fanny Kemble and many dignitaries were waiting on the platform to get on the train. The rain was coming from a spray bar from a road tanker parked on the road above. Of course, the water affected the ability of the brakes to stop the train, and judging the stop to get the door of the carriage to be exactly opposite the duke was virtually impossible. After 11 attempts and a whole tanker full of water, *Rocket* and the carriages finally stopped in the right place, by which time the locomotive crew had endured 11 drownings.

The second scene involved *Rocket* and its loaded train bursting out of Brooksbottom tunnel (half a mile long) at high speed with its complement of opening day guests. The driver then had to pretend to run down the unfortunate Mr Huskisson. As the photograph shows, there was less than six inches' clearance between the top of *Rocket*'s chimney and the roof of the tunnel. The train was required to accelerate to 30mph in the downhill section of the tunnel and burst out into the sunlight to get the perfect shot. The tunnel is on a curve, and forward vision was zero. Nobody had thought to ask the driver how long the stopping distance would be. It turned out to be nearly half a mile! *Rocket*'s unlucky volunteer fireman declared that he found the high-speed dash through the tunnel truly scary. He was amazed that National Railway Museum staff had such blind faith and courage to drive at such speed in the dark with so little braking power and no idea how far they would need to stop!

RIGHT Filming *The Day the World Took Off*. 'Are you sure it will fit through this tunnel?' *(Dave Campey)*

CHAPTER SEVEN

Epilogue:
the legacy of *Rocket*

Evening Star in 1960, the last steam locomotive
built for British Railways.

Introduction

The legacy of *Rocket* falls into five categories: industrial, social, cultural, technological/engineering and, for the author, personal. Did the Stephensons have any idea that *Rocket* would leave such a legacy? Do the creators of great and important engineering achievements ever realise that they have a world-beater on their hands at the time of its creation?

History judges winners and losers retrospectively and it is hard to recognise and critique great engineering achievements at the time of their creation. Did Henry Ford, for example, know that he had a transport world-beater in the Model T in the 1920s? Did Alec Issigonis know that the Austin/Morris Mini Minor he designed in the 1960s would become so loved and cherished by generations of drivers? I suspect not.

In the case of aircraft, iconic designs like the Tiger Moth, Spitfire or Lancaster have not faded into obscurity, but rather have become symbols of British engineering achievement; however, it is unlikely that their designers anticipated such a legacy. Even in modern times the project to get the last surviving airworthy Avro Vulcan bomber back to flying condition had never been seriously considered until 30 years after such bombers had been a regular sight in our skies. Yet when we see this great majestic bird in flight we cannot doubt that this surviving relic is a great machine that deserves to be cherished by younger generations.

How can it be that *Rocket*'s short life in its original form has had such a huge and significant influence, and why has replication of its iconic form taken on almost epidemic proportions? Clearly *Rocket* as a locomotive was not loved enough to be showered with accolades and feted once it had passed its usefulness after the opening of the L&MR. It was consigned to use as a test-bed for other motive power ideas and being a spare locomotive, because newer locomotives were more powerful and better suited to the task of running a busy railway. It was only by the efforts of James Thompson (described in chapter one) and a later generation of people at the Patent Office that the machine came to be regarded with wonder and affection and, miraculously, survived.

The industrial legacy

The success of *Rocket* at the Rainhill Trials led to the idea that commerce and industry could utilise the speed of railway locomotives to their advantage. Speedy intercity travel for business people became a reality. The idea was born that if inter-connected tracks were to spread throughout the country it would be possible to travel great distances swiftly, safely and cheaply by railway. The great change in the way that people and goods were transported by railway led to the gradual demise of canals and stagecoaches for everyday commerce and travel.

Nearly every one of the tens of thousands of steam locomotives produced in Britain owes its lineage to *Rocket*. Not every locomotive was built on Stephensonian principles, and there were several brave attempts to break the

BELOW *Ellerman Lines*, a Merchant Navy Class steam locomotive in the National Railway Museum's collection that was sectioned to show internal workings.

mould, but it feels strangely satisfying to me that the very last steam locomotive produced for British Railways in 1960 had the same three recognisable principal characteristics that *Rocket* combined to such good effect at the Rainhill Trials: a multi-tubular boiler, cylinders and pistons sloping down to directly power the driving wheels, and exhaust steam drawing the fire through a blast pipe under the chimney.

Similar innovations were introduced in the automotive industry, where designers tried to get away from the four or six cylinders in a line of motive power units by using the Wankel single multi-lobed rotor to replace pistons and connecting rods. However, many years on the cylinders-in-a-line version still reigns supreme.

The problem-solving processes that led to the genesis of different versions of *Rocket* by the Stephensons led directly to improvements in rail technology that made possible a viable locomotive-hauled railway system throughout Britain. The high speed and hill-climbing ability that *Rocket* demonstrated so effectively for the first time at the Rainhill Trials was a seminal step forward. As the steam locomotive improved over time, the dream of Britain as a nation linked up by a robust and rapid system for the transport of goods and passengers could at last become a reality. Coal, iron and stone, as well as other raw materials such as textiles, china clay, fish, meat and farm produce, could be shipped around the country to places where the finished goods could be most advantageously manufactured or marketed.

Although the Rainhill Trials proved that a steam locomotive was able to climb hills pulling a train of carriages behind it without the assistance of winding engines, there were terrains where the inclines were so steep that the wire rope hauling option could not be eliminated completely. When Robert Stephenson built the London and Birmingham Railway in 1837, the Act of Parliament for the railway specifically forbade the establishment of engineering facilities at Euston, so Robert Stephenson installed his engine shed and winding engine at Camden in an underground vault (which still exists) to drag the trains up out of the London terminus. Similar arrangements were put in place at Liverpool Lime Street and at Glasgow Queen Street. However, the

pace of steam locomotive developments was so great that rope haulage out of Euston was discontinued within six years of opening, though much later at both Liverpool and Glasgow. Nowadays rail passengers departing from those termini through dark tunnels in modern highly powered trains get no sense of the challenge that engineers had to overcome to transport previous generations of travellers and goods up and down those gradients.

One or two isolated pockets of rope-haulage survived right into the mid-20th century where the requirements could not be met by locomotives. One notable example was at Seaham Harbour in County Durham, where the heavily loaded downhill coal trucks were used to pull the empty uphill trucks back to the top without the use of any locomotives.

ABOVE Lithograph showing the chimneys of the steam winding engines installed at Camden for the climb out of London Euston station on the London and Birmingham Railway.

ABOVE Rope-hauled railway still in use 160 years after Stephenson's proof at Rainhill that steam locomotives could climb hills. This is the rope-worked incline at Seaham Harbour. (Andrew Scott)

BELOW Rope-hauled freight railway in 1952 between Cromford and Buxton in Derbyshire. The Cromford and High Peak Railway closed in 1967.

Another outpost of rope haulage until well into the 20th century was the Cromford and High Peak Railway in Derbyshire, which used a combination of wire rope and steam locomotive haulage to get over the Pennines between Cromford and Buxton until the closure of the goods-only line in 1967.

The success of Rocket at Rainhill put Robert Stephenson & Company at the forefront of steam locomotive building in Britain. As different countries round the world industrialised they needed railway systems to service their emerging industries, and the developers of rail systems beat a path to Stephenson's door to buy his locomotives. The world export market in steam locomotives initiated by the Stephensons' company sustained several locomotive building companies in Britain for well over a century. This period corresponded with the rise of the British Empire, during which, for well over a century, British-made goods were utilised and marketed throughout its colonies, to the benefit of British manufacturing companies. But the days of the steam locomotive were numbered, as cleaner, more efficient and less labour-intensive systems of providing railways with motive power were introduced. Electric trains had great advantages in countries like Switzerland and Norway, where hydro-electricity was readily available and coal was not. Diesel engine-hauled trains developed to replace steam in countries where oil industries thrived.

Early commercial steam locomotive-building companies were centred around established industrial areas in Britain: Beyer Peacock and Sharp Stewart in Manchester; Hunslet Engine Co, Manning Wardle, Kitson and Fowler in Leeds; North British and Neilson in Glasgow; Vulcan Foundry in Newton le Willows. In addition steam locomotives were built in large numbers by the individual railway companies, who commissioned engines from the commercial companies when demand outstripped their own manufacturing capability.

By the mid-20th century, after 130 years of supremacy on Britain's railways and in the export market, the writing was on the wall for the steam locomotive in Britain. The mechanical engineering expertise of the locomotive-building companies rallied behind British Railways' 1955 Modernisation Plan. There was a determination

to create modern and clean replacements for the outdated steam locomotives, and political kudos to be gained by those who could successfully and rapidly rid the country of its dirty, coal-burning, labour-intensive steam locomotives.

So, with inordinate haste, in the 1960s steam was replaced by largely unproven diesel, diesel electric and electric locomotives, often before the infrastructure to support such a radical overhaul was in place. The first generation of so-called 'modern traction' was being serviced in dirty and unsuitable steam depots under totally inappropriate conditions. Steam locomotives in their hundreds came to an ignominious end in railway scrap yards all over the country. Many of these locomotives were less than ten years old.

The folly of the undue haste with which the long reign of steam was terminated was demonstrated during the harsh winter of 1968. Teething troubles were only to be expected when brand new technology was flung straight into service. The sub-zero temperatures encountered during the severe weather of the 1960s meant that the diesel fuel powering many of the new generation of locomotives froze to jelly in the tanks. Their systems could not cope and there were examples of *stored* steam

locomotives being brought back into service from scrap lines to stand in for the failed new technology. How the Stephensons would have smiled at that!

Fortunately, by this time the British public's love affair with steam locomotives was becoming entrenched in the national psyche. It was sufficiently strongly felt for the government of the day to commit to saving for posterity a small proportion of significant steam locomotives from the cutters' torches. In 1975 this collection, of what was thought to be a representative selection of British-made steam locomotives, was combined with existing transport collections at York and Clapham to form the embryo of the National Railway Museum in the former steam locomotive depot at York. It was appropriate that one of the first national museums to come out of London-centric museum culture into the provinces was located in York, a natural railway centre as well as being located near the geographic centre of Britain.

The demise of the steam locomotive brought about a fascinating reversal of industrial development and entrepreneurship at an international level. In the 1960s the Japanese developed a high-speed electrically powered passenger railway called the *Shinkansen*,

ABOVE Barry scrap yard in South Wales in 1971, where steam locomotives were sent for disposal in large numbers at the end of the steam era.

ABOVE *Shinkansen* high-speed electric Japanese Bullet Train from the 1960s at the National Railway Museum.

between different versions of locomotive technology in different countries can be appreciated by the thousands of visitors from all over the world who visit the National Railway Museum. Here they can see iconic steam locomotives, including an enormous British-built Chinese railway engine, alongside diesel and electric-powered locomotives and the iconic *Shinkansen* train from Japan.

The social legacy

Britain's love affair with railways started in the heady days when George Stephenson drove *Rocket* proudly from Crown Street in Liverpool to Rainhill on the embryonic L&MR, hauling passenger coaches at speeds never experienced before. Ever since then, an affection for railways in general and trains in particular has remained a peculiarly British phenomenon.

As with ships, the naming of locomotives and trains has always been dear to the hearts of British people. All early locomotives had names. Was it just a pragmatic decision, so that an organisation that had more than one locomotive could be sure which machine was being referred to? I think not. Consider some of the names of the Stephensons' first batch of locomotives manufactured for the opening of the L&MR: *Rocket*, *Phoenix*, *Dart*, *Comet*, *Arrow*, *Meteor*, *North Star*. These names all suggested something spectacular and worth watching. Indeed, even the names of the Rainhill also-rans – *Sans Pareil*, *Novelty* and *Perseverance* – suggested something spectacular. One notable exception to the L&MR batch named above was *Northumbrian*. This name was Robert's respectful reference to his father's humble origins.

The early railways named their coaches also. The L&MR used *Experiment* and *Huskisson* as names for their coaches among others, rather like stagecoaches being named, but it was the locomotive names that caught on in a big way. Passenger locomotives tended to have names with which the travelling public could associate. Less glamorous freight locomotives tended to remain anonymous. The practice survives. In September 2015 a naming ceremony was held at the National Railway Museum for one of Virgin's High-Speed Trains, *National Railway Museum 40 years 1975–2015* (see photograph

worked by what have become popularly known in English as 'Bullet Trains'. This technology revolutionised rail transport in Japan, since when many other countries, like France, have followed suit. Speeds in excess of 180mph became commonplace with this new generation of trains.

In a gesture that signified how far the world's railways had progressed since steam power reigned supreme, in 2002 the Japanese railway authorities approached the National Railway Museum in York to see if the museum would be interested in accepting the donation of one of the first *Shinkansen* Series Zero trains, which were being withdrawn from service at the end of long and successful lives. In the 40 years of the Bullet Trains' domination of Japan's high-speed railway system, not a single fatality had been recorded. The Japanese logic was that 19th-century Britain had helped the world industrialise through the gift of the *Rocket* steam locomotive and its successors, but 20th-century Japan had created the high-speed electrically powered train that spread throughout the developed world.

The kind offer was taken up, and the links

at the bottom of page 93).

Apart from members of the general public who had affection for certain trains, there was that quirky band of locomotive enthusiasts whose passion was to see and *collect* the names and numbers of as many locomotives as possible. Seeing a locomotive meant it could be underlined in their notebooks.

There were, of course, locomotive enthusiasts in Britain before Ian Allan brought out his wonderful *Combined Volume* containing the numbers, details and statistics of all existing locomotives for such enthusiasts. The books offered data about classes of locomotives to accompany the lists of available numbers. The Ian Allan books brought train-spotting to the masses in the post-World War Two period.

The affection for railways did not stop with collecting numbers and names. *Rocket* is immortalised in one of the Thomas the Tank Engine books, called *Thomas visits the Great Railway Show* (at the National Railway Museum). And so a new generation of young children is treated to bedtime stories about the adventures of an odd-looking yellow, white and black locomotive with a very tall chimney and only one set of driving wheels!

The genesis of the inter-connected steam railway system in Britain enabled ordinary people to be more mobile. Not only could they consider commuting to work and so living in suburbia, but also the possibility of holidays became a reality through the advent of railway excursions. At the times of northern factory shutdowns or 'wakes weeks' the whole workforce and

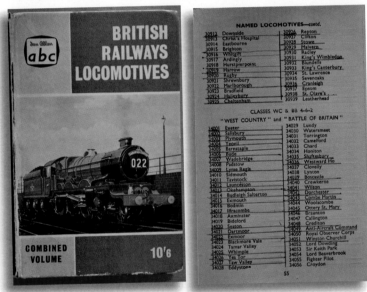

CENTRE Ian Allan's Combined Volume listed all the numbers of locomotives that train-spotters might expect to find across the country. Spotters would underline their various sightings to register their conquests! *(John Chambers)*

RIGHT Train-spotters at King's Cross in 1950. No doubt they all aspired to own an Ian Allan combined volume.

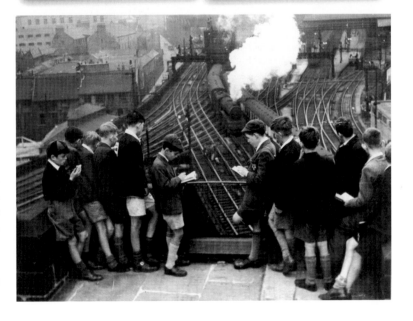

12 miles per hour for one penny per mile. Nowadays we have only first and standard class travel, whereas the railways of the 19th century catered for their travelling public with sometimes four classes, of varying degrees of comfort or discomfort. There were smoking and no-smoking compartments, as well as 'ladies only' compartments.

The cultural legacy

The British love affair with railways and trains is revealed through a range of cultural media leaving a lasting and enjoyable legacy. There were numerous films involving trains, but to me there have been none more classic, poignant and enduring than the film *Brief Encounter*, made in 1945 by David Lean, where the coming and going of trains at a busy railway junction brings together, then parts for ever, two unhappy people. The trains both facilitate, and end, their relationship.

In the film of *The Railway Children*, a dramatisation of the Edith Nesbit novel, we are again treated to a heart-wrenching story in which trains act as a conduit to convey important messages to the children, whose absent father has been falsely accused of spying. It may seems tenuous to connect *Rocket* with *Brief Encounter* and *The Railway Children*, but if we revisit the Fanny Kemble

ABOVE The author trying to read the number of a passing excursion train at Warrington Bank Quay station. *(Author)*

BELOW Southbound excursion trains lined up at York Station in 1910 ready to travel to holiday destinations.

their families might travel en masse to holiday destinations like Llandudno and Scarborough for a week beside the seaside.

In 1844 the government of the day legislated for the railways to run what they called Parliamentary Trains, which stipulated cheap fares for the travelling public. At least one train per day had to be run on every British Railway line in each direction. These Parliamentary Trains had very basic facilities, with nothing more than seats and a roof. They had to run at speeds of not less than

journals discussed in chapter three we can perhaps understand the connection between trains and romance, and see that she was swept off her feet by the glamour of the railway and the people that the train introduced into her life.

Railways and especially the steam train make popular subjects for artists, and perhaps none is more famous than J.M.W. Turner's painting *Rain, Steam and Speed*, which represents the excitement of the steam train only 15 years after *Rocket*'s success at Rainhill. In the 20th century Terence Cuneo's paintings were able to take us back through time to recreate the thrill of great railway moments – for example, his *Opening of the Stockton and Darlington Railway* captures vividly how this event was seen and experienced at the time.

Art became a part of public rail travel in the early days as railway posters on station billboards gave passengers the chance to dream of places to which they could be taken by train. In order to tempt them to travel there, passengers in carriage compartments were also treated to eye-level artistic interpretations and photographs of the places to which a particular railway company was able to take them. Railway companies commissioned their own artists to create these carriage prints and publicity posters to entice the public to use their company's railways more.

Within railway poetry we find ourselves fortunate that some poets' love of the railway became enshrined in their work. Examples include John Betjeman's *Pershore Station*, Edward Thomas' *Adelstrop*, Wilfred Owen's *The Send-Off*, Thomas Hardy's *On the Departure Platform*, Philip Larkin's *Whitsun Wedding*, and many others. From Berlioz to Butterworth, music and railways reflect the idiosyncratic rhythm that trains create when travelling over rail joints.

W.H. Auden's 1935 *Night Mail* combines film and poetry to create the rhythmic and glamorous excitement of how our postal service worked through the night to bring mail to our doorsteps.

Perhaps the most famous is *Coronation Scot*, written in 1937 to celebrate the LMS's launch of their powerful new Coronation class

of locomotives, matched in their launch publicity material with the replica *Rocket*, as described in chapter five.

Are these special artistic outpourings in different genres something that is rooted in the steam locomotive specifically, or are they likely to continue to emerge from the modern railway? Only time will tell.

ABOVE Terence Cuneo's painting of the opening of the Stockton and Darlington Railway in 1825.
(Cuneo Fine Arts)

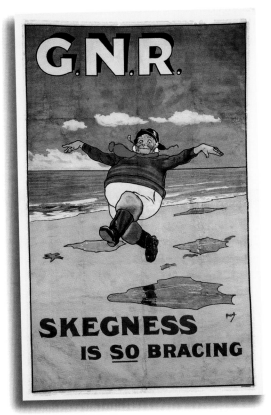

LEFT 'Skegness is so bracing'. One of the many holiday posters commissioned and issued by the railway companies to encourage the public to travel with them.

The technological and engineering legacy

Perhaps the greatest engineering legacy to come from *Rocket* was Robert Stephenson himself. Following the success at Rainhill his company went from strength to strength. Robert went on to undertake several prestigious railway-engineering projects. Whilst the Newcastle-based company continued to concentrate on locomotive building, Robert followed his civil engineering leanings. His greatest works that survive include the London and Birmingham Railway, Kilsby tunnel, Newcastle's High Level Bridge over the Tyne, Conway tubular railway bridge and the remains of his Menai tubular bridge, as well as the fabulous Border bridge at Berwick-upon-Tweed.

What of the Stephensons' legacy as inventors? There has always been controversy over the blast pipe arrangement of *Rocket* and whether it was invented by Timothy Hackworth. The point is that the Stephensons did not claim to be inventors – they were more like entrepreneurs who were happy to incorporate innovative ideas and best engineering practice into their locomotives, from any source. Robert Stephenson certainly experimented with the blast arrangements to get the most effective draught-inducing arrangements for *Rocket*, but others had used the system before. He altered the initial *Rocket* blast pipes into a single unit to offer less restriction to the exiting flue gases than the two individual nozzles. We should not be surprised by this kind of experimentation and development. The science behind the proportioning of the chimney and blast pipe arrangements in the smokebox of a steam locomotive was something of an empirical black art that was not mastered until nearly at the end of steam locomotion, when, in the 1950s, draughting expert Sam Ell at Swindon Works finally nailed the ideal proportions and formula for optimum effect of the blast pipe and chimney.

The Stephensons' use of a multi-tube arrangement to increase the heat exchange between fire and water was to have far-reaching implications for the industrial world, beyond railways. Within ten years of the Rainhill Trials the static traditional twin-flued Lancashire steam boiler was being transformed by placing a multi-tubular stack of tubes called an 'economiser' into the path of the flue gases on their way to the chimney. The feed water was pumped through these tubes on its way to the boiler. The result was an immediate improvement of 25% in thermal efficiency, with corresponding fuel savings. Multi-tubular heat exchangers became the norm for applications where compactness was important. Today we see such heat exchangers installed in applications from motorbikes and cars to power stations and blast furnaces.

The Stephensons chose to make their external water-jacketed firebox out of copper for two reasons. Firstly, the thermal conductivity (the relative ability to conduct heat rapidly through the metal) of copper is some five times that of iron and steel. Secondly, copper is a much more ductile material and could be more easily stretched and formed into the complex double-lined horseshoe shape required for *Rocket*'s boiler. The down side was that the softness of the copper meant that its flat surfaces had to be supported by stays in many more areas than the same shape in iron, but the improvement in heat transfer ability and therefore the efficiency of the machine was spectacular. The lasting legacy of the decision to use copper for the first time in a locomotive firebox was that many railway locomotives were built with copper fireboxes right to the end of steam locomotive building in Britain in the 1960s.

A personal legacy

I have been privileged to have been involved during my working life with both the original *Rocket* (through the 'dig' with Dr Michael Bailey and Dr John Glithero described in chapter one) and by being able to take the 1979 replica *Rocket* to various places around the UK and the world to demonstrate it as part of my role as chief engineer at the National Railway Museum. An important benefit of demonstrating such a significant machine, even though it is a replica, is that it forces whoever is looking after it to get into the head of its creators: Robert Stephenson, his father and their colleagues. For example, I have described how we gained new insights into the challenges they faced in providing a competent reversing system and drive to the valves. It took until 1843 to get a foolproof reversing mechanism established on British locomotives.

It might seem to the reader that *Rocket* is just a scaled-down version of later locomotives, and that if you know how to drive a full-size locomotive you will naturally be able to drive *Rocket*. Happily some of us have had the opportunity to do both. Driving *Rocket* is a unique challenge. We have covered earlier in the book the idiosyncratic reversing mechanism and technique used to drive *Rocket*, which contributed to the untimely death of William Huskisson on the opening day of the L&MR in 1830. In order to drive *Rocket* competently drivers have to understand the valve-shifting mechanism and the purpose for which it is employed. A close analogy from the automotive world might be the technique known as 'double de-clutching' to facilitate gear changing on elderly cars and lorries with 'crash' gearboxes. Unless you fully understand the process, it is near impossible to execute the technique competently and achieve a smooth gear-change. This dilemma has led to some delightful exchanges when the *Rocket* replica has visited some heritage railways where a roster of volunteer drivers is often strictly adhered to. For such loan agreements the National Railway Museum's representative who brings a locomotive to the site normally has no role other than overseeing the safety and sensitivity to the locomotive on loan. A rostered driver for the *Rocket* replica might have had a lifetime of driving full-size steam locomotives, but faced with *Rocket*'s controls he might as well have sat in the pilot's seat of a helicopter for the first time and been expected to fly it!

One of the delights of being involved with steam locomotives like *Rocket* and its successors is working with a machine that so completely interacts with those charged with its care. We now live in an age when we expect purchased goods to have a CE mark or a British Standards Kitemark to inform the user that it has been manufactured to an agreed standard, and therefore can be expected to perform to an agreed specification. Every car, steam iron or umbrella that comes off a factory production line is guaranteed to perform in a certain way, independently of the skill of the user. The manufacturers pride themselves on this achievement and use it in their advertising material.

That is definitely *not* the case with a steam locomotive, and certainly not with *Rocket*. A big influence over the machine's ability to perform satisfactorily is the way that the driver and fireman interact with the machine, and I am so grateful that I have lived in an age when I have had the privilege to master and enjoy driving steam engines and steam locomotives. I think that of all the machines I have been involved with throughout my life *Rocket* is the quirkiest!

The power output from a steam locomotive is difficult to define. Although there is a nominal figure derived from the energy given by the fuel and the weight of the machine upon the rails, there are so many variables. These can allow short bursts of power well in excess of the textbook-rated power. This is partly what endears so many people to the steam locomotive. It is one of the few machines that can respond really well to being driven hard under the right conditions. It is, of course, a self-sustaining device. By using different driving or firing techniques, and paying careful attention to water levels in the boiler, each steam locomotive can assume an independent character. In fact I could swear that some days a steam locomotive is even capable of *sulking*. In contrast, how dull it is that every diesel or electric locomotive off the production line nowadays performs exactly the same as its fellow classmates, however it is handled!

I can think of only two other examples, alongside the steam locomotive, of devices created by mankind where the behaviour from a machine is totally dependent on the skill of the person operating it: a sailing boat and a weaving loom. The interaction between the operator and the device determines what happens and is dependent on the skill, knowledge and experience of the person operating it.

To return to the idiosyncrasies of *Rocket*, although we have the remains of the original *Rocket* and there have been many non-working replicas, the real value in terms of public understanding and knowledge comes from having the original *and* the working replica available to observe and understand them together.

My own special moments with the 1979 replica *Rocket* are numerous. They range from having the *Rocket* replica in steam operating whilst jacked up off the ground on the ramparts of Norwich Castle, to performing for the crowds on the shores of Lake Lucerne at the Verkehrshaus museum in Switzerland. At times it has felt like having a personal magic carpet! But two personal highlights concern the making of the BBC *Timewatch* programme that recreated the Rainhill Trials, detailed in chapter six. Firstly, it gave the engineering team of the National Railway Museum the chance to implement a radical solution to alter the *Novelty* replica so that it became fully operational for the first time. Secondly, we had a chance to restage the Rainhill Trials in 2002 in such a way that, in spite of everything, *Rocket* was still the outright winner. This brought our National Railway Museum team enormous personal satisfaction and pride. It was particularly poignant for me when, after my retirement from the National Railway Museum, I was invited to Gävle in Sweden to take part in a festival celebrating the steam locomotive and was able to get involved with the replicas of *Rocket* and *Novelty* competing once again with each other.

There is one final remarkable twist in the story of my personal legacy of *Rocket*, which occurred after I had retired from the National Railway Museum and largely lost professional contact with *Rocket* and her replicas. It potentially brings together the two great personalities of Robert Stephenson and Isambard Kingdom Brunel, who were good friends and supported each other at difficult times in their respective careers.

I was asked in 2008 to get involved with a project to complete the installation of replica steam engines inside the SS *Great Britain*, the world's first great iron ocean liner, built by Brunel in 1843. At the time of writing it is on display in the museum at Bristol. I was employed as 'steam guru' to the main contractor, Heritage Engineering of Glasgow.

Replica engines were constructed that work through an electric motor and inverter drive. The engines had to be less than a tenth of the weight of the originals due to the weak state of the hull. Although the components looked like very heavy castings they had to be designed to be really light and manageable. The original 1843 engines were strongly influenced by Isambard Kingdom Brunel, who designed and built the ship itself. The four huge double-acting cylinders with 88in bore and 72in stroke were designed so that it was possible to run each pair of cylinders in steam in different directions. Although this sounds counter-intuitive it was a very convenient way of manoeuvring the great ship at low speed, and as a bonus gave almost instantaneous reversing. Of course, only two cylinders were delivering power to the propeller at any one time in this condition.

The question of how to reverse the steam engines when working as a coupled four-cylinder unit must have taxed Brunel considerably. There is a challenge with any ship's engine connected to paddles or a propeller. Assuming the ship is travelling forwards under power and needs to reverse, steam is shut off; but the engine continues to turn whilst the propeller drives it. It is therefore necessary to get the engine into reverse whilst the ship and the engine are still travelling forwards. If this sounds similar to the technical section described in chapter two, on how *Rocket* reverses, then read on.

All the time whilst I was drawing out the reversing gear to be manufactured in Glasgow, from the scant information available, I became convinced that somehow the sliding clutch mechanism was strangely familiar, but could not quite figure out why. Eventually the penny dropped. I realised that what was shown on John Weale's contemporary engraved drawings of the engines of SS *Great Britain*, that I was trying to draw out in a lightweight version, was the slip eccentric cluster from the front axle of *Rocket* with which I had become so familiar. But the mechanism was ten times *Rocket*'s size! It was a light bulb moment – I found myself speculating whether Brunel might have been discussing with his friend Robert Stephenson the difficulties of reversing the ship under power. Maybe Robert suggested trying the solution he had created for the front axle of *Rocket*. There are so many similarities in the mechanism. As with *Rocket*, if the shaft is not turning in the SS *Great Britain* engine when the engineer wants to reverse, then it is hard work using the large barring hand gear wheels to turn the engine until the correct forward or reverse dog engages in the drive plate. Surely I cannot be the only person to have made the connection between these great and historic engineers' work?

Maybe the reason why the link has not been noted previously is that marine engineers might feel they had little in common with steam locomotive engineers. Maybe nobody

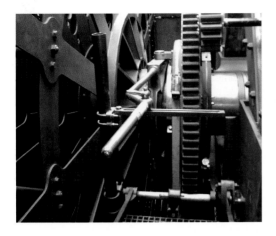

ABOVE Reversing eccentric cluster used on the replica steam engines installed in the SS *Great Britain* at Bristol in 2005. *(Author)*

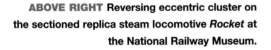

ABOVE RIGHT Reversing eccentric cluster on the sectioned replica steam locomotive *Rocket* at the National Railway Museum.

had crossed those disciplines in heritage engineering until now? Either way, the shapes involved are too similar for it to be mere coincidence. I like to imagine these two great men mulling their respective engineering problems over and perhaps sketching on the equivalent of beer-mats, until they came to the solution we see now in action on both replicas. If this were true what a delightful synergy between two great engineering minds. The world's first proper steam locomotive and the first proper steamship perhaps share a secret? Now *that* is special!

There is a splendid quote in L.T.C. Rolt's book *George and Robert Stephenson*, published in 1960: 'Robert Stephenson, Brunel and Locke died before their time. Their triumph was also their tragedy. With their deaths a great era of heroic endeavour drew to a close. For good or ill they laid the foundations of the modern world.'

Half a century later Stephenson's *Rocket* retains its rightful place as the centrepiece of the 'Making the Modern World' exhibition in the Science Museum in the heart of London.

That is its legacy.

RIGHT What adventures await the remains of Stephenson's *Rocket* in the next hundred years?

Appendices

Ready for the *Timewatch* Trial. *Rocket* replica and National Railway Museum team wait to start their loaded test run on the delightful Llangollen Railway during *Timewatch* retrials.
(Martyn Stevens)

Appendix 1

The significance of the Dr Michael Bailey and Dr John Glithero report

The key findings of Bailey and Glithero's report were:

- *Rocket*'s working life was only three to four years, after which it was deemed not powerful enough to run the daily service. The process of continual improvement in the newer locomotives coming out of the Stephenson factory in Newcastle meant that *Rocket* was no longer the preferred motive power for traffic between Liverpool and Manchester.
- Once that working life was over, *Rocket* remained out of service until 1862 apart from a brief revival on the Earl of Carlisle's colliery railway under James Thompson.
- *Rocket* was involved in four separate accidents during its working life. The damage was sufficient for the locomotive to be either repaired in the workshops at Edge Hill or, in some cases, sent back to the factory in Newcastle.
- The out-of-square copper firebox from the copper-bottoming factory of James Leishman and John Welsh in Liverpool meant that the errors had to be accommodated by adjustments to the locomotive's frames in the earliest stages of *Rocket*'s build.

BELOW Copper firebox delivered out-of-square to Robert Stephenson & Co in 1829. *(John Glithero)*

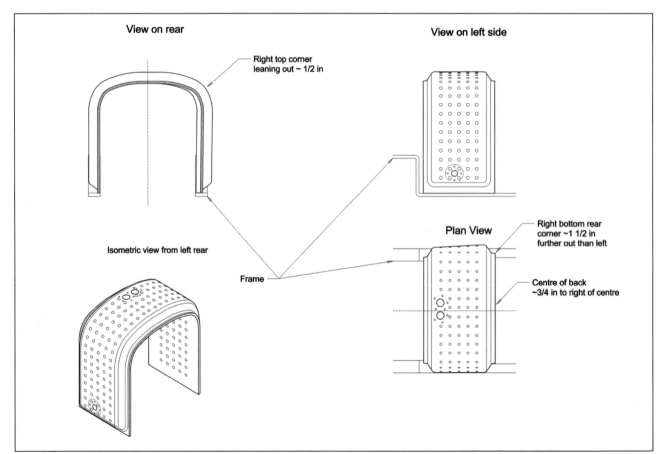

The myth about the bulged shape of the inside surface of the copper firebox was finally resolved. John Rastrick's notebook sketches were vindicated as accurate. Of course, this meant that the Robert Stephenson centenary replicas, as well as the Mike Satow replica, were based on incorrect interpretation and continue to tell an incorrect story.

Rocket was probably the first locomotive to have slip-eccentric reversing with dog clutches to select the valve timing by sliding the eccentric cluster from side to side.

Rocket was originally built without a dome on the top of the boiler. A dome is a device which enables pure steam to be drawn from the boiler, *ie* excluding water: this is ironic in view of the fact that all the Rainhill Trials replicas were built *with* domes! What looks like a dome on the front cover of this book is in fact the lock-up safety valve sitting on top of the manhole cover that gave access to the interior of the boiler. Once the dome was installed in 1830 and the safety valve relocated further back, it enabled the

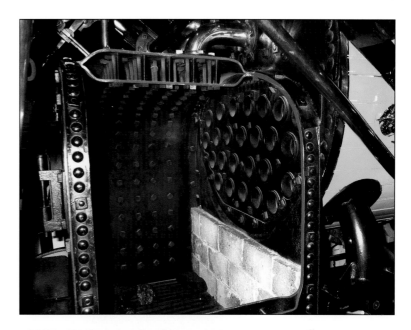

ABOVE 1935 *Rocket* replica firebox with incorrectly shaped inner bulge in the sectioned water jacket.

BELOW Eccentric cluster with dog clutches sliding side to side to engage with either forward or reverse drive plates to reverse the locomotive. *(John Glithero)*

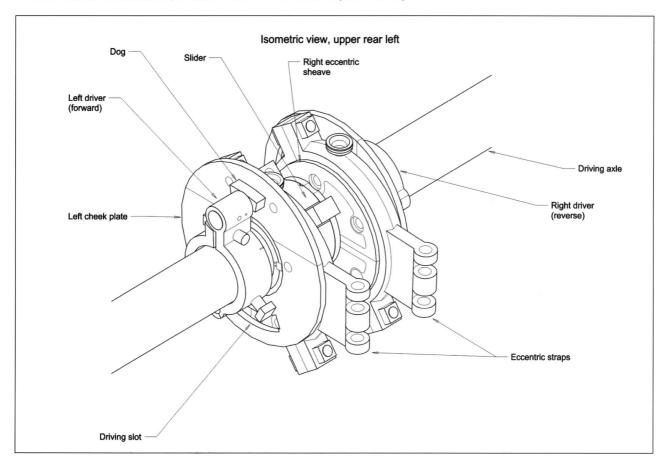

Isometric view, upper rear left

Dog

Slider

Right eccentric sheave

Left driver (forward)

Driving axle

Right driver (reverse)

Left cheek plate

Eccentric straps

Driving slot

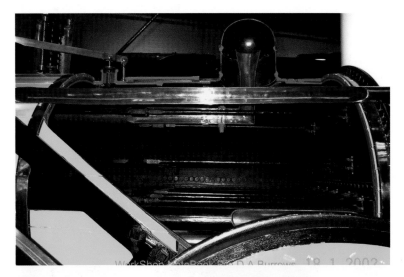

steam to be collected from an inverted pipe positioned inside the dome. This in turn allowed the water level in the boiler to be raised by 3in, improving the thermal capacity of the boiler and reducing the tendency for boiler water to be carried over into the cylinders when *Rocket* was driven hard (priming).

■ After the Rainhill Trials the blast pipe was changed from two separate nozzles to one combined nozzle. This improved the capacity of the locomotive to sustain its combustion by the exhaust steam being blasted up the chimney drawing in fresh air under the grate. The argument was that a combined blast nozzle would restrict the chimney less than two individual nozzles.

■ *Rocket*'s front axle was increased to 4in diameter from 3in during one of the rebuilding programmes following an accident. This probably would have entailed fitting new horn-blocks and axle boxes. It would have been done to make the front axle, valve gear and wheel fixings more substantial in view of damage sustained during previous accidents.

■ The replaced driving axle shows signs of having been badly grooved by the gear driving plate fastened to the axle. The head of the locking bolt scored the axle deeply.

TOP Replica boiler with dome as *Rocket* was eventually modified, but incorrect for the 1829 version.

ABOVE Replica with separate blast pipes as originally built in 1829. *(Author)*

RIGHT Supplementary frames and buffing gear added to allow *Rocket* to couple at the front.
(John Glithero)

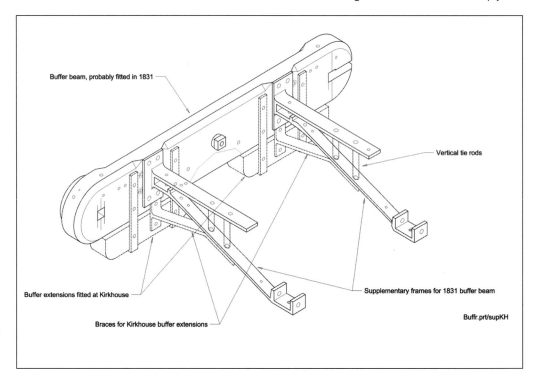

Buffer beam, probably fitted in 1831

Vertical tie rods

Buffer extensions fitted at Kirkhouse

Supplementary frames for 1831 buffer beam

Braces for Kirkhouse buffer extensions

Buffr.prt/supKH

- *Rocket*'s cylinder mountings were changed from side to side when the cylinders were lowered from 38° to 8° to the horizontal following rebuilding.
- The right-hand-side driving wheel was replaced by a different one at some point, possibly as a result of one of the derailments.
- Supplementary front frames and buffers were added to allow *Rocket* to couple at the front to rolling stock.
- The smokebox underwent a series of modifications and enlargements as the original was found to be under-capacity for the quantity of ash generated in a day's work. The third and present smokebox was fitted in 1836.
- Erroneous replica components were fitted in 1862 to try and create a museum display that more truly represented what visitors expected. This process leads us right to the heart of the debate about what information can be passed on when an artefact is displayed in its final rather than its initial condition.
- A firebox 'wet back' system was installed to get more heat out of the firebox and dispense with the refractory firebox back-plate. The remnants of that system still exist.
- On many of the L&MR's early locomotives an ash pan was fitted to lessen the risk of lineside fires as red-hot cinders from the fire dropped on to the track. Fitting the ash pan to the base of the firebox with a damper control meant that by restricting the air opening to the underside of the grate, the fireman could have more control over the rate of fuel burning and so could more carefully control steam production and optimise combustion. There is no evidence of *Rocket* having been fitted with an ash pan in its short working life on the L&MR.
- In 1834 *Rocket* was used to trial the unsuccessful rotary engine of Lord Dundonald.

RIGHT Remains of 'wet back' firebox of *Rocket* drawn in final form. (*John Glithero*)

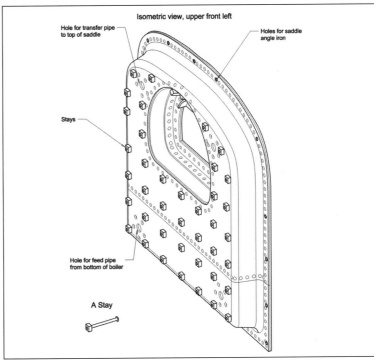

Isometric view, upper front left

Hole for transfer pipe to top of saddle

Holes for saddle angle iron

Stays

Hole for feed pipe from bottom of boiler

A Stay

Appendix 2

Glossary of some terms used in this manual

Bar frames – Locomotive frames made out of iron or steel bar to create a strong cradle to support the boiler, wheels and cylinders of a locomotive. The bar frames carry the pulling and pushing forces involved, and keep those forces away from the boiler. *Rocket* has bar frames but most British locomotives have plate frames, which are two deep plates rigidly fastened together, with openings across which the axles fit.

Cast iron – A material cheaply produced in a blast furnace from iron ore, coke and limestone. It is strong in compression but weak in bending and tension, and tends to break like glass. Chemically it is pure iron with 5% free graphite mixed in with the iron. The flakes cause weakness when overloaded in tension or bending. It is not good for railway rails.

Coke – A clean, hot smokeless fuel derived from burning coal under limited oxygen supply in a closed vessel to remove all the gases and volatile matter. It was the chosen fuel for steam locomotives like *Rocket* but gave off noxious fumes when burned in a locomotive firebox. It was used in large quantities to melt iron in blast furnaces.

Crank – An eccentrically mounted drive pin connected to the driving wheels that enabled reciprocating motion from the piston driven to and fro by steam to be converted into rotary motion, as with a pedal car.

Crosshead – The pivoting joint between the piston rod and connecting rod usually guided to travel back and forth in a straight line allowing the force from the piston rod to reach the crank, so turning the wheels.

Cylinder drain cocks – When hot steam first enters a cold cylinder and tries to push the piston the steam condenses to water, which impedes the flow of steam. It is necessary to drain the condensed water away from the bottom of the cylinder via cylinder drain cocks. Once the cylinder is hot, the cocks can be shut to minimise the wastage.

Disposal – The railway term for taking a steam locomotive that has finished its day's work and preparing it so that it can be left safe and ready for work the next day.

Dog clutch – A dog clutch couples together two rotating components with a peg that fits firmly into a hole. There is no gentle sliding of the components before they are brought together as with a car clutch. It is all or nothing; but the components always come together in the same position.

Double-acting – The pistons inside the cylinder of a steam engine are called 'double-acting' if the crosshead is both pushed and pulled by the steam in the cylinder acting on either side of the piston in turn.

Draughting – The fire burning in the firebox is drawn by the action of the exhaust steam exhausting up the chimney and causing a draught of air through the grate. The proportions of the blast pipe in the smokebox and chimney base are critical to make the fire draw properly.

Eccentric – A disc of iron mounted on a

BELOW *Rocket* replica rests after its exertions at the *Timewatch* Trials. The evidence of hard prolonged running shows on the chimney base.

rotating axle. The hole is offset from the centre so that when the eccentric rotates the strap around the outside imparts a reciprocating action for driving a valve or a pump.

Flange – A raised rim of metal around the circumference of a railway wheel that engages with the inside of the rail and stops the pair of wheels falling sideways off the track.

Gab – A means of uncoupling the fixed valve gear of a steam locomotive so that the valve it drives can be worked independently of the eccentric on the front axle.

Gland – Another name for a stuffing box. It is a way of sealing a sliding shaft into a cylinder when the cylinder has steam pressure inside it.

Keyway – A rectangular slot along a shaft where the component fitted on the shaft must be prevented from rotating but can still slide along the shaft as needed. A rectangular key fits into the slot and keeps both components rotating together.

Multi-tubular – By using many small tubes through the middle of a vessel rather than one large tube, the heat transfer between the medium passing through the tubes and the medium outside the tubes is hugely increased.

Puddling – The process of turning molten cast iron into pure wrought iron by manually stirring it in a furnace in an atmosphere that carries away the free graphite, leaving pure iron. It is a batch process that is technically challenging. The finished product, 'wrought iron' or 'malleable iron', is strong in compression, tension and bending and is ideal for making strong railway lines. It can only be made in limited quantities. Wrought iron produced by puddling was largely replaced by steel after Henry Bessemer discovered how to make steel easily in 1854.

Priming – A locomotive is said to be 'priming' when the water level in the boiler gets too high and the water is carried over into the cylinders with the steam. Incredibly the water defies gravity and climbs up out of the free surface into the steam pipe, which causes the locomotive to slow and become unmanageable.

Regulator – Also known as a throttle. The regulator controls the flow of steam from the boiler to the two cylinders of *Rocket*, regulating the amount of steam being fed to the cylinders and thus determining the power output of the locomotive.

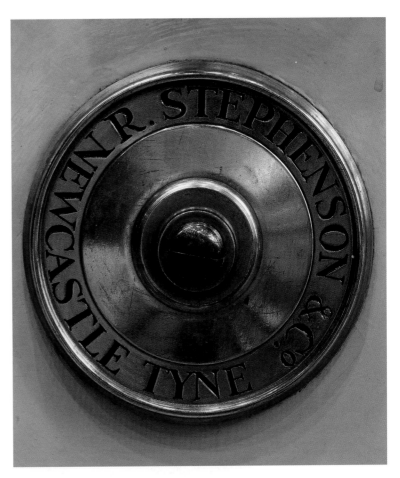

ABOVE Replica driving wheel centre with maker's plate.

Slide valve – A plate, omega-shaped in section, which slides within the valve chest directing the steam and the exhaust into the passages where it is needed. The valve is driven to and fro by the valve gear.

Slide bars – These guide the crosshead in a straight line so that the out-of-line forces transmitted by the connecting rod do not disturb the straight-line movement of the piston rod and crosshead.

Stuffing box – See *gland*.

Turnbuckle – An adjusting device fitted in the middle of an iron rod. It has left- and right-hand threads that, without dismantling, allow the length between the two ends to be easily varied whilst under load. They are ideal for boiler stay adjustment.

Wet back – A wet-back system was installed to water cool the firebox doorplate and improve efficiency. The remnants of that system still exist.

Wrought iron – A flexible, strong, corrosion-resistant metal that is ideal for structural use. Long used in small quantities for armour, nails, screws and chains. See *puddling*.

Appendix 3

Rocket's working life timeline

(Courtesy of Dr Michael Bailey and Dr John Glithero)

1829	
March	Directors of L&MR announce Trials and prize of £500.
April to August	Manufacture and fitting at R. Stephenson & Co.
September	Trials and shipment to Liverpool.
October	Rainhill Trials and demonstration trips.
November to December	Engineering trains, Chat Moss contract.

RIGHT Detail of *Rocket*'s left-hand driving wheel, spring and nameplate.

1830	
January to July	Engineering trains, Chat Moss contract.
August	Demonstration runs Liverpool to Manchester.
September	Maintenance, opening day, engineering trains.
October to November	Engineering trains, accident 28 October, repairs.
December	Main line duties.
1831	
January	Main line duties.
February	Repairs and modifications.
March to July	Main line duties.
August to December	Engineering trains and track repair work.
1832	
January to August	Engineering trains and track repair work.
September to October	Loan to Wigan branch railway.
November	Accident 6 November and repairs.
December	Repairs.
1833	
January	Repairs.
February to August	Secondary duties and standby locomotive.
September	Used for motive power experiments.
October to December	Standby locomotive.
1834	
January to September	Standby locomotive, infrequently used.
October to November	Dundonald rotary engine trials, fitting and removal.
December	Out of use.
1835	
January to December	Stored out of use.
1836	
January to May	Stored out of use.
June to July	Restored for sale.
August to September	Available for sale.
October	Sold to the Earl of Carlisle.
November	Shipped to Kirkhouse.
December	Repairs by R. Stephenson & Co.
1837	
January to February	Repairs by R. Stephenson & Co.
March to July	Colliery train duties.
August	Alston election run, 8 August.
September to December	Colliery train duties.
1838	
January to May	Colliery train duties.
June	Acquired by James Thompson.
July to December	Colliery train duties.
1839	
January to December	Colliery train duties.
1840	
January to December	Withdrawn from service and preserved.
1862	Donated to Patent Office Museum.
1884	Department of Science & Art (South Kensington Museum).
1909 to date	Science Museum Group collection.

Appendix 4

Relevant places to visit

Science Museum

A major museum on Exhibition Road in South Kensington, London. It was founded in 1857 and today is one of the city's major tourist attractions. The extant remains of *Rocket* are on display in the museum's 'Making the Modern World' exhibition.

Address Exhibition Road, London, SW7 2DD
Telephone 0870 870 4868

Museum of Science and Industry, Manchester

A large museum devoted to the development of science, technology and industry, with emphasis on the city's achievements in these fields. The original eastern terminus of the Liverpool and Manchester Railway is part of the museum.

Address Liverpool Road, Manchester, M3 4FP
Telephone 0161 832 2244

BELOW *Rocket* replica leaves the National Railway Museum under tight tunnel clearance on Leeman Road to adventure to pastures new. Where I wonder?

National Railway Museum

This museum in York forms part of the Science Museum Group of National Museums and tells the story of rail transport in Britain and its impact on society.

Address Leeman Road, York, YO26 4XJ
Telephone 0844 815 3139

Locomotion: The National Railway Museum

The National Railway Museum at Shildon, where the original *Sans Pareil* is on display.

Address 'Locomotion', Shildon, County Durham, DL4 2RE
Telephone 01388 777999

Beamish Museum

The Pockerley Waggonway, early replica locomotives and a replica engine house can be seen at this north of England open-air museum located at Beamish, near the town of Stanley, County Durham.

Address Regional Resource Centre, Beamish, County Durham, DH9 0RG
Telephone 0191 370 4000

George Stephenson's birthplace

This cottage at Wylam, Northumberland, is run by the National Trust.

Address George Stephenson's Birthplace, Wylam, Northumberland, NE41 8BP
Telephone 01661 843276

Rainhill Community Library

Rainhill was the site of the 1829 Rainhill Trials won by George Stephenson and *Rocket*. The Trials exhibition is staged in a British Railways Mark 1 carriage in the grounds of St Helens Council's Rainhill Library. It contains sketches and accounts of the Trials with original memorabilia. A short audiovisual presentation traces the development of the railways and steam locomotives. The Rainhill Trials display is maintained by the Rainhill Railway and Heritage Society.

Address View Road, Rainhill, Prescot, Merseyside, L35 0LE
Telephone 01744 677822

Cromford and High Peak Trail, Middleton Top Engine House and Sheep Pasture incline

The winding engine at Middleton top is operated every first weekend between April and October, plus Bank Holiday weekends in summer. It is today operated by compressed air rather than steam.

Address Middleton Top Countryside Centre, Middleton by Wirksworth, Derbyshire, DE4 4LS
Telephone 01629 823204

Head of Steam: Darlington Railway Museum

Head of Steam, formerly known as the Darlington Railway Centre and Museum, is located on the 1825 route of the Stockton and Darlington Railway, which was the world's first steam-powered passenger line. Stephenson's 1825 *Locomotion* is on display.

Address North Road Station, Darlington, County Durham, DL3 6ST
Telephone 01325 460532

Kelham Island Museum, Sheffield

This industrial museum alongside the River Don in the centre of Sheffield has Stephenson's original link motion steam engine on display.

Address Alma Street, Sheffield, S3 8RY
Telephone 0114 272 2106

Swedish Railway Museum

Family-friendly railway museum with old locomotives (many early British) that you can climb inside.

Address Rälsgatan 1, 802 91 Gävle, Sweden
Telephone +46 10 123 21 00

ABOVE Rocket replica heads for home after completing the Swedish adventure. *(Tom Sandstedt, Swedish Railway Museum)*

Index

Accidents 32, 34
Aircraft, iconic designs 130
Allan, Ian 135
Anti-railway lobby 74

Bailey Dr Michael 7-8, 12, 34, 37, 53, 64-65, 82,
    85, 95, 99-100, 103-104, 106, 116-117, 120-
    121, 138, 144-147
Barry scrap yard, South Wales 133
Beamish Museum 18, 105, 110-111, 120, 153
    Pockerley Waggonway 110-111, 153
Bedlington Iron Works 17
Beyer Peacock, Manchester 132
Bird, Charlie 87
Birkinshaw, John 17
Blast pipe 9, 131, 138
Blenkinsop, John 14
Boiler tubes 19, 106
    clinking 58-59, 75
    copper 20, 23-24, 58-59, 104
    economiser 138
    iron 20
Boilers and related systems 14, 16, 18-21, 113-115
    axle mountings 23, 46
    expansion 23, 49
    explosions 62
    horizontal return flues 113
    hydraulic testing 23-24, 59, 115, 126
    Lancashire 138
    multi-tubular 9, 18, 20, 22-23, 29, 44, 58,
        105-106, 120, 131, 138, 149
    pressure gauges 47
    single flue tube 18-19, 28, 106, 110-111
    stays 24
    try cocks 60, 113
    two fire tubes 18, 138
    vertical 112
    water gauges 60, 78
    water levels 60, 139
Bolton & Leigh Railway 40
Books and journals
    abc Brtiish Railways Locomotives Combined
        Volume (Ian Allan) 135
    England in 1815 by Joseph Ballard 72
    George and Robert Stephenson by L.T.C. Rlt 141
    Record of a Girlhood by Fanny Kemble 1883 72-74
    Story f the life of George Stephenson by Samuel
        Smiles 1862 74
    The Engineering and History of 'Rocket',
        Science Museum 2000 8, 37
    The Royal Magazine 1909 78
    Thomas visits the Great Railway Show 135
    Tracts on Mathematical and Philosophical
        Subjects by Charles Hutton 11
Booth, Henry 18-19, 22-25, 29, 56, 58, 69, 106
Bowes Railway 18
    Springwell workshops 105
Braithwaite and Ericsson 26, 28, 41, 107, 115
Braithwaite John 26, 28-29, 41, 115, 120
Brakes 32, 44
Brandreth, Thomas 26, 117
Bridgewater, Duke of 79-80
British Railways 8-9, 131
    1955 Modernisation Plan 132
Brunel, Isambard Kingdom 39, 140-141
    Being Brunel exhibition, Bristol 39
    drawing instruments 39
Brunel, Marc 39
Brussels Exhibition 1910 59
Burrows, Dave 95
Burstall, Timothy 24-25, 28

Campey, David 126-127
Canals 130
    Bridgewater 72
Canterbury & Whitstable Railway 16
Cars, iconic models 130
Channel 4 TV The Day the World Took Off 126-127
Chapman & Buddle 110
Chimneys 20, 131, 138
Clay Cross Company 56
Clinker 93, 118
Coalbrookdale, Shropshire 9
Cochrane, Thomas (Lord Dundonald) 34, 147
Coppull Colliery, Chorley 40
Cowlairs Incline 48
Cranked axles 34
Creevey MP, Thomas 72-74
Cromford and High Peak Railway 132, 153
    Middleton Top Engine House 153
    Sheep Pasture Incline 153
Cylinders 14, 24, 34, 40-41, 50-51, 55, 131
    direct drive 9
    pistons 54, 131, 148
        double-acting 14, 54, 148
    vertical 110-111

Darlington Bank Top station 14
Darlington North Road Railway Museum (Head of
    Steam) 14, 111, 153
Davidson, Peter 116
Diesel locomotives 132-133, 139
Driving and firing positions 66-68

Earl of Carlisle's colliery 34, 144
East Lancashire Railway 126
    Bolton Street station 127
    Brooksbottom Tunnel 127
Electric trains and locomotives 132-134, 139
Ell, Sam 138
Ellis, Matt 124
Entwhistle, Edward 78-81
Ericsson, John 26, 28, 41, 115, 120

Films 136
    Brief Encounter 73, 136
    Old Hospitality 102
    The Railway Children 73, 136
Fireboxes 31, 68, 93
    ash pans 147
    copper 123, 138, 144
    fire-bars 65, 93, 118
    grate 64-65, 93, 1118
    separate (external) 20, 22-23, 138
    stays 23, 138
    twin 19
Flour Mill workshops, Forest of Dean 49, 106,
    121-122
Footplate 67-68
    cooking breakfast 124
Ford, Henry 98, 101-103, 106, 122, 130
Fowler, Leeds 132
Frames 48-49
    bar 23, 48, 148
    plate 48
    wooden 110
    wrought iron 121
Friction pairs 91
Friends of the National Railway Museum 113
Fuel and water supply 68

Gibbon, Richard 6, 120, 126, 136, 138-141
Gland packing removal tools 92

Glands on sliding joints 91-93
    graphite-lubricate yarn 91, 93
Glasgow Queen Street station 48, 131
Glithero, John 7-8, 12, 23, 53, 64-65, 82, 85, 95,
    99-100, 103-104, 106, 111, 116, 120-121, 138,
    144-147
Great Exhibition 1851 35, 100
Great Northern Railway 22

Hackworth, Timothy 17, 24-25, 27-28, 62-63, 66,
    68, 107, 138
Hedley, William 14, 17, 110
Heritage Engineering, Glasgow 140
Horse haulage 9, 16-17, 26, 29
Howe, William 56
Hunslet Engine Co, Leeds 132
Hurrell, Andrew 115
Huskisson MP, William 32, 81, 127
    Memorial, Parkside 34
    struck by Rocket 32-33, 139

Inclines 15-16, 31, 77-78, 131-132, 153
Industrial espionage 24-25
Issigonis, Alec 130

Japanese Bullet Train (Shinkansen) 6, 74, 133-134

Kelham Island Museum, Sheffield 56, 153
Kemble, Fanny 72-74, 79, 81, 127, 136
Kennedy, John 27, 58
Killingworth Colliery and waggonway 10
King's Cross station 135
Kitson, Leeds 132

Lamb, Richard 11
Lean, David 136
Leishman, James 144
Little Eaton Gangway (Derby Canal Railway) 16
Liverpool & Manchester Railway (L&MR) 6, 14 et
    seq.
    absorbed by L&NWR 98, 101
    Chat Moss bog 10, 30-31
    construction 72, 80
    directors 31, 72, 77-78
    Edge Hill workshops 38, 144
    gradients and inclines 15, 31, 77-78
    Liverpool Crown Street station, Edge Hill 15, 31,
        74, 77-79, 134
        Moorish Arch 33, 79
    Manchester Liverpool Road station 152
    named coaches 134-135
    Olive Mount cutting 77, 80
    opening 29, 31-34, 73, 78, 80-81, 94, 99, 126,
        130, 139
        cavalcade 32
        ticket 32
        water stops 33
    Parkside station 32-33
    Rainhill 6, 116
        October 1829 exhibition of Rocket 72
        skew bridge 25, 70
    replica coaches 66, 135
    route map 15
    Whiston Plane 77-78
    Wigan branch 34
Liverpool Lime Street station 48, 131
Liverpool Mercury 72
Llangollen Railway 116
Locke, Joseph 32, 141
Locomotion, NRM at Shildon 40, 152
Locomotion Enterprises 105, 121

Locomotive fitters 91, 93, 95
Locomotives – see also *Rocket*
  Agenoria 8
  Arrow 134
  Coalbrookdale 1803 (first built) 9
  Comet 134
  Cycloped 26-27, 29
  Cycloped replica 117
  Dart 134
  Duchess/Coronation class 101-102, 137
  Ellerman Lines (sectioned 'Merchant Navy' class) 130
  Evening Star (last built for BR) 8-9, 28, 81, 128, 131
  Flying Scotsman 6, 120
  Lancashire Witch 18-20, 22, 50, 65
  Locomotion 14-15, 17-22, 52, 58, 65-67, 106, 111, 153
  Locomotion replica 18, 105, 110-111
  Losh and Stephenson design 1826 15, 17
  Meteor 134
  National Railway Museum 40 Years 1975-2015 (Virgin HST power car) 134
  North Star 134
  Northumbrian 30-32, 51, 134
  Novelty ('London Engine') 26-27, 29, 31, 40, 65-66, 69, 120, 134
  Novelty replica 107, 111-120
    preparing to run 112-115, 140
    wheel extensions 113, 115, 120
  Novelty non-working replica 40-41
  Pen y Darren 8
  Pen y Darren replica 110
  Perseverance 26-28, 134
  Phoenix 134
  Planet 34
  Puffing Billy 14, 17, 58, 110
  Puffing Billy replica 110-111
  Royal George 17
  Salamanca 14
  Sans Pareil 8, 24-25, 27-28, 31, 40, 65-66, 68, 134, 152
  Sans Pareil replica 67, 107, 110-111, 116, 118
  Steam Elephant 110
  Steam Elephant replica 110-111
London and Birmingham Railway 131
  Camden Bank 131
  London Euston station 48, 131
London Midland & Scottish Railway 102, 137
London & North Western Railway (L&NWR) 49, 59, 63, 98, 101-102
  Crewe Works 99-100

Manchester Museum of Science & Industry 40-41, 152
Manning Wardle, Leeds 132
Manpower Services Commission 105
Mechanics' Magazine 29
Middleton Railway, Leeds 14-15
Midland Railway Centre 81
Miniature steam locomotives 54
Models
  Novelty 112
  penny-in-the-slot *Rocket* 100
Moore, Alan 103, 122
Morris, Graham 120-122
Mosley, David 81, 103, 125
Motion 54-55
  connecting rods 110
  piston rods 110
  slide valve 54, 149
Museum of British Transport, Clapham 102, 105, 133
Museum of Rail and Track, Angelholm, Sweden 111, 113, 120
Music 137

National Collection 14, 40, 100, 111, 122
National Railway Museum (NRM), York 6-7, 28, 34, 62, 81, 98, 100, 103, 105, 111, 126, 130, 133-135, 152
  archives 23, 55, 69, 101
  BBC gift of *Cycloped* replica 117
  insurance company 122

launch of Virgin Super Voyager 93
store 52-53
workshops 85, 112
yard 95
Neilson, Glasgow 132
Nesbit, Edith 136
North British, Glasgow 132

Paintings and drawings depicted
  conjectural drawing 1859 by J.D. Wardale 36, 100
  Crewe drawing of Rocket 63
  George Stephenson 10
  lithograph of Camden winding engines 131
  Opening of the Stockton & Darlington Railway by Terence Cuneo 137
  Rocket at Rainhill by Alan Fearnley 70-71
  The Collier 1814 14
Parker, Bill 121-122
Parliamentary Trains 136
Patent Office, London 35-36, 40, 130
Penydarren, South Wales 9
Perkin, Deborah 111
Phelps, Geoff 122-123
Poetry 137
Posters 21, 35, 65-66, 137
Priming 64, 146, 149

Rails 17
  Birkinshaw 17, 70
  cast iron 16, 18
  fish-bellied 15, 17
  wrought (malleable) iron 17, 70
Rainhill Community Library 40-41, 153
Rainhill Levels 15, 31, 77-78
Rainhill Railway and Heritage Society 153
Rainhill Trials, 1829 9, 21-23, 25-29, 48, 65, 69, 98-99, 131
  also-rans 40-41, 134
  centenary celebrations 101-102, 104
  competing contestants 27-28
  documents 24
  judges 25, 27, 58
  locomotive tenders 65-66
  original entries 26-27
  performances 27
  prize 27-29
  public attendance 25, 27-29, 69
  rules and conditions 25, 29, 47, 52, 58-59, 72
  150th anniversary – see *Rocket 150*
Rainhill Trials re-enactment 2002 72, 106-107, 111-120, 140
  BBC *Timewatch* programme 106-107, 111-120, 140
  judges 116-118
  results 118-119
Rastrick, John 23, 27, 53, 102, 104, 106, 145
  sketches and notes 58, 60, 62, 87-88, 99, 104-106
Rees, Jim 103, 106, 110, 120-122
Regulator 64, 149
Reversing mechanisms 54-56, 139-140
  slip eccentrics 54-55
Robert Stephenson & Co. 7, 10, 21, 24, 34-36, 99, 132, 138, 140
  Darlington works 65, 104
  Forth Street Works, Newcastle 15, 21-22, 29, 31, 35, 38, 52, 99, 111, 138, 144
  ledgers 38
  presentation book for Henry Ford 101

***Rocket* - original locomotive**
  accidents and repairs 29, 32, 34, 38, 49, 85, 144, 146-147
  archaeological examination (the 'dig') 7-8, 12, 36-38, 64, 82, 94, 103, 106, 115, 120-121, 138, 144-147
  'Rocket Round-up' 37-39, 115
  axle 38, 52-55, 140, 146-147
  axle boxes 53, 75, 90, 146
  bearing adjustment 90-91

blast pipes 9, 29, 57
boiler 9, 18, 22-24, 27, 44, 46, 56-60, 81, 88-89, 138
  barrel (wrought iron) 56, 88
  dome 44, 64, 145
  fastened to the frame 48-49
  end tube plates 24, 57-59
  hydraulic testing 23-24, 59
  manhole 89, 145
  pressure 47, 59, 76-77, 80-81
  solid deposit build-up 89
  stays 24, 59
  try cocks 60
  tube space 56-57
  tubes 19, 56, 58-59, 86, 89
  washouts 89
  water levels 75-76, 104, 146
braking 78
chimney 29, 36, 45-47, 51, 56-57, 69, 76, 86, 146
crosshead 45, 50, 61, 91
  water pump 61-62, 76-77, 92, 94
construction 21-22, 39
couplings and buffing gear 31, 66
  carriage draw-gear 66
  front of loco 31, 147
  with tender 66, 68
cylinders 9, 11, 44-45, 50-54, 61, 76, 78, 91-92, 99
  drain valves 75-76
  lowered 11, 29, 49-50, 53, 99, 147
  mounting on frame 49
    90° out of phase 54
  pistons 22, 50, 53, 61
  stuffing boxes (glands) 51, 91-93
  swapped from side to side 53, 147
drivers and firemen 72-81
  evacuated to Brocket Hall 36
exhaust 115
exhaust pipes 36, 51
firebox 22, 23, 44, 46, 56-58, 75, 86-87, 93, 99-100, 104-107, 121, 144
  bulge myth 104-106, 145
  distortion 23, 144
  fire-bars 65, 75, 77, 93
  fire-hole door 75, 77-78
  grate 64, 93
  lighting the fire 75
  stays 58
  wet back system 147
fake photograph 99
footplate 67-68, 72, 75-76
  travelling on 72-79
frame 23, 38, 46, 48-49, 66, 68, 144
  supplementary 147
horn blocks and guides 53, 146
lubrication 53, 75, 78, 85, 89-91
maintenance 85, 89-91, 95
manometer 47, 76
modifications and updates 31, 37, 85, 147
motion 54-56, 69
  connecting rods 45, 91
  piston rods 51, 92
  slide bars 50
  valve gear 51, 56, 76, 146
museum exhibit 11, 35, 121
  donated to Patent Office/Science Museum 35-36
  erroneous replica components fitted 36, 147
  loan to Merseyside County Museum 36
  NRM, York 7, 12, 36-39
  Second World War evacuation from London 36
nameplate 7
paintwork 68-69
performance 27, 131
  power 102
  speed 27
  tractive effort 48
  Rainhill Trials condition 6, 44, 50
  vital statistics 44
Rainhill Trials winner 29, 31, 99, 132, 153

rebuilds 52, 147
regulator 45, 57, 76-77
  valve plug 64
remains 11, 27, 31, 57, 69, 99, 139, 141
reversing gear 38, 68, 76, 138, 140
  slip eccentric cluster 38, 55-56, 68-69, 140, 145
riding behind 72, 78
safety valves 62, 64, 89, 145
smokebox 11, 29, 44, 57, 86, 147
sold by L&MR 35
soot and ash deposits 86
  cleaning 86-87
springs 22, 46, 53
steam chest 50-51, 76
storage 7, 34-35, 144
tender 61, 65-66, 68, 75, 93-94
  sherry (water) barrel 61, 65, 75, 93-94
used for experiments 34, 147
valve chest 50, 92
valves 54, 69, 138
visit to Japan 6, 36
water circulating pipes 57, 89
weight 22, 24, 53, 58
wheels 7, 22, 43, 45, 52, 61, 69, 94, 147
  bearings 75
  daily inspection 95
  rear 52-53
  shrunk-on tyres 52, 94
  strengthening 22-23, 43, 52
working life 29, 31, 37, 94, 144, 147, 150-151

Rocket replicas 97-127, 145
  all known 98
  1881 Crewe replica 98-102, 105, 121
  1923 Buster Keaton movie replica 102
  1929 Robert Stephenson & Co. 98, 100-101,
  103-104, 106, 122, 145
    bulged firebox 104-106, 145
  1937 rebuilt Crewe replica 102

Rocket 1935 sectioned replica 28, 65-66, 98,
  102
  blast pipes 46, 146
  boiler 57, 146
    adjustable stays 60
    clinked tube 59
    dome 63, 146
    try cocks 60
    tube space 57
    water level gauge 60
  chimney 46-47
  connecting rod 45, 91
    big end 45
  coupling height 66
  crankpin 91
  crosshead 45, 61, 91
  cylinder 45-46
  debut at the Albert Memorial, London 124
  exhaust pipe 46
  firebox 57, 106
    bulge 145
    grate 64
  footplate reversing pedal 68
  mercurial manometer 47
  piston-rod 45, 92
  regulator 64
  reversing eccentric cluster 141
  safety valves 62
  valve-rod 92
  water circulation pipes 88
  water feed pump 61
  water gauge 86
  wheels 45
    flanges 94
    loose wooden spokes 95
    rotating crank 45
    wear 95

Rocket 1979 and 2010 working replicas 7, 11,
  62, 69, 72, 102-103, 107, 116-127
  air brake system 81
  ash pan 95, 118
  boilers 81, 89, 105-106, 120-123
    front tubeplate 84
    inspections 84
    penetrated by screws 125-126
    pressure 81
    scaling 89
    tubes 81
    washout 88
    wooden cladding 125
  braking system 122, 127
  buffer beam 7
  building the 2010 replica 120-124
  chimney 86, 105-106, 127
  derailment 95
  driving and firing 72, 74, 81, 103, 126-127, 139
  fireboxes 81, 88, 105-107, 120-121, 123
    bulge 106, 145
    fire-hole door 87
    grate 93
    stays 123
  fire lighting 106
  fitness to run examination 85
  fixed cut-off 81
  frames 49, 120-123
  lubrication chart 84-85
    using modern oils 85
  maintenance 84-86, 95
  out-of-the-ordinary events 95, 124-127
    steaming at Norwich Castle 139
  performance 85
  repainting 125
  reversing 68, 81, 103, 139
    footplate pedal 81
  runs in re-enactment Rainhill Trials 112, 116-119
  runs in Rocket 150 cavalcade 7, 53, 107
  safety valves 62
  soot and ash deposits 86-87
    cleaning 87
  stopping 81, 127
  tender 93-94
    water shut-off valve 94
  valves 81
  visit to Sweden 119-120, 140
  visits overseas 120-121, 140
  water gauge maintenance 85-86
  wheels 52-53
    rear 95, 120-121

Rocket 150 anniversary celebrations 1979 6-7, 36,
  53, 70 , 107

Safety valves 62-63
St Helens & Runcorn Gap Railway 40
Satow, Mike 6, 52, 102-103, 105-106, 110-111,
  120-121, 145
Science Museum, South Kensington, London 6, 27,
  31, 35-36, 40, 49, 52, 99, 103, 112, 121, 152
  Land Transport Gallery 36
  Making of the Modern World exhibition 7, 11, 14,
    37, 39, 110, 141, 152
Seaham Harbour 131-132
Sharp Stewart, Manchester 132
Ship's engines 54, 140
Sjoo, Robert 111, 119
Smokeboxes 31, 34, 138
South Kensington Museum 35, 79
Springs 48
  leaf 22
  steam 15
SS *Great Britain* 39, 140-141
Stagecoaches 68, 72, 74, 78-79, 130, 134
Stationary steam winding engines 9, 15-16, 20, 28,
  31, 77-78, 131

colliery engines 18, 54
  rope haulage 48, 131-132
Steam, live and exhaust 20, 54, 57, 131
Stephenson (née Henderson), Frances 10
Stephenson, George 8 *et seq.*
  birthplace 153
  death 99
  engineer for Canterbury & Whitstable Railway 16
  engineer and surveyor of L&MR 15
  engineer for Stockton & Darlington Railway 99
  five-pound note 30-31
  Geordie brogue and charm 10, 73, 76
  'Geordie' miners' lamp 10
  marriage 10
  single parent 10
Stephenson, Robert 8 *et seq.*
  apprenticeship 11
  bridges and tunnels 11, 138
  engineer for London & Birmingham Railway 131,
    138
  study at Edinburgh University 11
Stirling, Patrick 22
Stockton & Darlington Railway 14-15, 17-18, 52 99,
  111, 153
  150th anniversary 105, 111
Swedish Railway Museum, Gävle 119-120, 153
Swindon Works 138

Tenders 65-66
Thomas the Tank Engine 6, 135
Thompson, James 34-35, 130, 144
Track systems 16-18
  modern 112
  poorly maintained point-work 52, 95
  rack and pinion 14
Tractive effort 47-48
Travel classes 136
Trevithick, Richard 8-9, 20

Valve gear 55, 66
  gab action 55, 149
  Stephenson's link motion 56, 153
Vulcan Foundry, Newton-le-Willows 132

Wallsend Waggonway 110
Wardale, J.D. 36, 100
Warrington Bank Quay station 136
Water quality and treatment 86-89
Watt, James 54
Weale, John 140
Wellington, PM Duke of 32-33, 73, 79,
  127
Welsh, John 144
Weston, Dame Margaret 121
Wheels 14-15, 22, 34, 47, 94-95
  cast steel spoked 52-53
  chilled cast iron 53
  couplings 22
    chain 15
    side rods 22
  fastened to boiler 23, 46
  metal 23, 94
  spoked bicycle 112
  two-piece plug 52
  tyres 94
  wear 94-95
  wooden 22, 52, 94-95
  90° out of phase 48, 52, 54
Whessoe Ltd 106, 120
Wood, Nicholas 11, 27, 58
Wood, Walter 78
Wordsell, Thomas and Nathaniel 65
Worsted wool trimmings 75, 90
Wylam Colliery 14-15, 17, 110

York Railway Museum 133
York station 136